W9-AEA-080

RELIGION, CIVIL SOCIETY AND THE STATE

Configuration of religion, civil society and the state as depicted in a miniature Mughal illustration of the Jahangir period, 1605–27. It shows a young prince being received by a faqir seated on a platform in the wilderness, a known theme also of the *Book of kings* (*Shahnamah*). Here the young prince, the infant state, is shown as standing in a modest attitude while being presented by his nurse or nanny, who is unveiled, and may be taken as representing civil society rather than the family. The city whence this party hails has been superadded in the top background.

Bodleian Library, Oxford, MS Ouseley add. 175, ND

RELIGION, CIVIL SOCIETY AND THE STATE

A Study of Sikhism

J. P. S. UBEROI

DELHI
OXFORD UNIVERSITY PRESS
BOMBAY CALCUTTA MADRAS
1996

Oxford University Press, Walton Street, Oxford OX2 6DP

Oxford New York
Athens Auckland Bangkok Bombay
Calcutta Cape Town Dar es Salaam Delhi
Florence Hong Kong Istanbul Karachi
Kuala Lumpur Madras Madrid Melbourne
Mexico City Nairobi Paris Singapore
Taipei Tokyo Toronto
and associates in
Berlin Ibadan

ISBN 0 19 563691 0

Typeset by Rastrixi, New Delhi 110070
Printed in India at Pauls Press, New Delhi 110020
and published by Neil O'Brien, Oxford University Press
YMCA Library Building, Jai Singh Road, New Delhi 110001

Preface

The Indianness of India, from a national, structural or relational view, will be found to be inversely proportional to its Hinduness or Hindutva, for example, as this is understood by its chief exponent in the twentieth century. V.D. Savarkar, who hated Gandhi from the time that they knew one another in London, 1908, wrote that the strength, cohesion and progress of India depended, in the last resort, upon the strength of Hindutva, relegating questions of social reform and Hindu–Muslim unity to mere side issues (1923: 119–28). The view that he canvassed at length is entirely European and modernist, not to say Orientalist, in origin. It identifies the strength of any nation or people with the principle of homogeneity or uniformity, the coincidence of its land, race, language, culture, religion, etc., culminating in the state. As against this is the pluralistic view that the unity of humanity and of its several segments, secular or religious, depends rather upon the non-coincidence or overlapping of the component races, languages, cultures, classes, etc., which produces or can produce unity through equality and difference, relations of mutuality, complementarity and exchange in society, 'unity in variety'.

Even if, by some process, all non-Hindus were to be extruded from the 'heart of Aryavrata' and its dependencies, they would remain somewhere next door, and the problem of the relation of self and other, or rather of self and the other self, would still remain to be addressed. This problem of humanity, national and international, cannot be solved within a framework of the majority and the minority, superordination and subordination, the centre and the periphery, the mainstream and the margin or homogeneity and heterogeneity. It can be solved only by the reconciliation or negotiation of equality and difference, competition and co-operation, the convergence of underlying structures or the distribution of differences into complementary domains as the bases of mutuality, reciprocity and exchange, whether in amity or enmity.

Savarkar, whether or not he was the inspiration of terrorists and assassins, brought all his talent and energy to focus on the project expressed by the slogan that he coined in 1941, 'Hinduize all politics and militarize Hindudom!' He was a hero of the national freedom movement; and for him the Hindus in themselves constituted 'the foundation, the bedrock, the reserved forces of the Indian state'. Gandhi, who led the movement in civil society, was the martyr of *swaraj*, supreme witness to the causes of Hindu–Muslim unity, removal of untouchability, and of *swadeshi* as the love of one's neighbour and of his or her labour. For all the lip service paid to him after the event, and also to his legacy of method, the theory of *satyagraha* or putting the truth of God and society above the state, Gandhi now lies more or less rejected in India. Savarkar saw the strong Hindu state as being effective in offence—rather than only in defence—and capable of dictating its terms to the world regarded as a system of states or of nation-states, with scarcely a thought for society or reform, let alone for pluralism and Hindu–Muslim unity. Gandhi wanted the 'equality of all religions' (1930) as the foundation of modern Indian society, nationality and culture, whether organized as one independent state or two (1947), and he consistently refused to subordinate means to ends, either for himself and his own or in relation to any other segment of humanity, which he regarded as a system of national and international societies, peoples and cultures. Needless to say, the method and praxis of *satyagraha* or Gandhian non-violence Savarkar rejected and declared to be suicidal as well as sinful (Dhananjay Keer 1950: 219ff).

The difference between the two leaders, from the standpoint of the present study, lies equally in their corresponding attitudes to European or Western modernity. Savarkar's father was a landowner known for his Western-style education as well as his Sanskrit scholarship, and the son brought equal disgrace to both traditions. He and his 'New India' group, fascinated by the prospect of politico-military action as the way of national salvation (*dharmayuddha*), went and secretly learned the art of bomb making from a failed Russian revolutionary in Paris. Indeed, this strain of Hindutva, because it is fearful of dialogue, will always want to combine India and Europe, tradition and modernity, in a non-relation of pure metonymy, perfect dualism and without any mutual conversation or any sense of reflexivity or self-criticism applied to either point of reference. In other words, it must always implicitly depend, like all projects of political communalism or revivalism—Hindu, Muslim or Sikh—upon an imported modernity, since it is incapable of producing any principle of motion from within itself.

Gandhi, on the contrary, would always look the other in the eye as his other self and offer conversation without fear of possible consequences. For him the national freedom movement or *swaraj* meant essentially the self-rule and self-reform of society, and Ram Rajya was to bring the rule of religion-in-society, a 'Kingdom of Heaven' in politics viewed as self-management of civil society rather than of the state. While telling Tagore as well as the British that 'our non-cooperation is a retirement within ourselves' (1921), he in fact produced out of tradition a modernity of India's own—so that India could come into its own in the world with a reformed tradition of pluralism. As we shall see, this was also the moment of perfect collaboration between non-cooperation as the form of the national freedom movement, on the one hand, and the self-reform of Sikhism, 1920–25, on the other, by the methods of *satyagraha*, martyrdom and non-violence.

The same attempted combination of pure anti-Muslim Hindutva, passed off as tradition, and a juxtaposed mindless pro-Western modernity, already finds expression in the Bengal renaissance of the century before. Its greatest luminary of the late nineteenth century was Bankim Chatterjee, 1838–94, who both presents and explains the mystery in the last chapter of his most important work, the novel *Anandamath*. Here the two intersecting dualisms of fact and value, i.e. 'physical' and 'spiritual', and of theory versus practice, i.e. 'knowledge' and 'action', which constitute the specific matrix of European modernity, as I have argued elsewhere (*Science and culture*, Delhi: Oxford University Press, 1978), are fully internalized and presumed by Bankim to be self-evident universals of truth and reality. And the looked-for saviour is again the state, and not any change of civil society, no matter that the state must be a foreign body, as in speaking of the *sannyasi* revolt, *c.* 1770.

Your task is accomplished. The Muslim power is destroyed. . . . What is going to happen now is for the best. . . . True Hinduism consists in knowledge, not in action. Knowledge is of two kinds, physical [fact] and spiritual [value]. Spiritual knowledge is the essential part of Hinduism. If however physical knowledge does not come first, spiritual knowledge will never comprehend the subtle spirit within. Now physical knowledge has long since disappeared from our land, and so true religion has gone too. If you wish to restore true religion, you must first teach this physical knowledge. Such knowledge is unknown in this country because there is no one to teach it. So we must learn it from foreigners. The English are wise in this knowledge and they are good teachers. Therefore we must make the English rule. Once the people of India have acquired knowledge of the physical world from the English, they will be able to comprehend the nature of the spiritual. There will then be no obstacle to the true Faith (*Anandamath* or *The Abbey of Bliss*, 4: 8, transl. T.W. Clark *in* de Bary 1958: 715f).

The life of the sacred, civil or moral society in Islam, I hasten to add, was equally lost in the period of medievalism somewhere between the priest and the prince, an hypotrophied tradition, on the one hand, and an hypertrophied state, or at any rate its executive branch, on the other hand. We shall have to determine in this study first the elementary structure of this unfortunate dualist condition, Hindu and Muslim, before one can return with confidence to the problematic of the Indian modernity of religion–in–society and the non-dualism of unity in variety.

The chief difficulty about introducing Sikhism in this universe of discourse as the new modern form of non-dualism, religion–in–society, is the same as will arise with the later example of Gandhism. It is due to an unacknowledged and unexpressed principle of closure, a bar which is widely assumed, whether consciously or unconsciously. Around AD 1200, if we turn to Europe, it was understood that there was to be no new religion or revelation, e.g. the second coming or like the tenth *avatar*, and any novelty of religion was to be classified, therefore, as schism, heresy, religious protest or reform. The Christian Pope, Innocent III, relying upon both imperial and canonical tradition in his decretal, *vergentis in senium*, addressed to the clergy and the people of Viterbo, for the first time identified heresy with treason, i.e. from the view of the established single religion of the priest and the prince, between whom lay the potential space of civil society. After 1208, the spiritual sword of dialogue was finally cast aside, and papists and statists were united in the view that 'iron was to conquer those whom persuasion would not convince', thus combining religion and politics in the worst possible way and to the utter detriment of civil society.

In the classical or medieval view that is still held by many, then, the doors of revelation and interpretation were closed by authority long ago in history, and no one can now seek to establish in the world a (new) society for self-realization. F. Schuon, a respected and learned student of comparative religion, having confessed his adherence to this standpoint of the closing of the gates (say) after Islam, or in general the middle ages, then vainly tries to make a sole exception in favour of Sikhism, as an example of the 'very latest possibility', and that also made possible by the quite exceptional context, as he says, of the proximity of Sufism and Hinduism (1963: 50n). But I hold and would urge to the contrary that the co-existence, dialogue and exchange of aspects among religions are typical of humanity and civil society, and by no means exceptions in history as distinct from historians.

For it is a law of religion and society, and not only for the Christian martyrs, that whosoever tries to save his self shall surely lose it, but the one who will lose himself for the sake of truth, God, other human beings or nature, shall find his true self in love of the other. I cannot imagine why we ever needed the British to divide and weaken our society, while pretending to strengthen the state, when the prevalence of the one true religion, say, of the khaki shorts and the saffron flags or Hindutva, as pitched against the one true religion of some other Indian, usually Muslim, can do it so much better and simultaneously leave the modernity of science, technology and economy dependent upon foreign tutelage, inspiration and control or leadership. Let us put the praise of national pluralism in the form of four propositions:

(*a*) that civil society is truly the locus of usage or custom, not of tradition; (*b*) it is the sovereign arbiter of custom as against the priest, custodian of tradition, and the prince, maker or executor of the state and its law; (*c*) inspired by religion or secularism or, as I think, pluralism, civil society alone has the inherent power to change usage, custom and itself. That is the modern and Gandhian realization of *swaraj* or self-rule and self-reform in civil society, established by custom and common usage in the space opened between tradition and the state or the priest and the prince, and it is surely the good way to combine religion and politics in the management of institutions. The Hindus, whether or not they were given the name by the Muslims and their sacred Vedas recovered for them in the written form by Europeans who were Orientalists, are no more (or less) qualified in themselves than non-Hindus to speak for India under the regime of pluralism; and whether or not the Sikhs are to be classed as Hindus is a purely subordinate question that we shall not attempt to answer. (*d*) The priest and the prince, whenever they rule together, either through a state-established religion or a religion-established state, are the enemies of civil society, its national autonomy, customs and morality.

I should not have ventured into the field of social and cultural studies of India at all but for the welcoming kindness of two men: Niharranjan Ray, the historian, who was the first director of the Indian Institute of Advanced Study, Simla, who is now alas no more; and M.N. Srinivas, the sociologist, who inducted me in 1968 into the department of sociology, Delhi School of Economics, University of Delhi, where today I recall specially Veena Dua and M.S.A. Rao, Vidyasagar Reddy and Rami Sanwal, who have all gone prematurely before their time. Acknowledgments are due for a variety of help to numerous other friends, officials, colleagues and enemies who gave shape to my project and life, but I cannot name all of them. H.S. Gill, formerly of the Punjabi University, Patiala, now professor of linguistics, Jawaharlal Nehru University, New Delhi, taught me semiotics as well as Sikhism. W.R. Roff, last at Columbia University, New York, who follows a different approach, nevertheless kept a best friend's vigil through all travails of the study of Islam first at the

Australian National University, Canberra, and then at Monash University, Melbourne. This venture of mine is dedicated to all the students of sociology at the Delhi School of Economics with whom it was my privilege and fortune to come into contact—and to Patricia Uberoi and Anuradha Shah, first and last of the line of pupils who taught me.

Delhi School of Economics J.S.U.

Contents

Plates

Figures

On the Jewish Question

Religion is precisely the recognition of man in a roundabout way, through an *intermediary*. The state is the intermediary between man and man's freedom. Just as Christ is the intermediary to whom man transfers the burden of all his divinity, all his *religious constraint*, so the state is the intermediary to whom man transfers all his non-divinity and all his *human unconstraint*.

Man, as the adherent of a *particular* religion, finds himself in conflict with his citizenship and with other men as members of the community. This conflict reduces itself to the *secular* division between the *political* state and *civil society*. For man as a *bourgeois*, "life in the state" is "only a semblance or a temporary exception to the essential and the rule". . . . The difference between the religious man and the citizen is the difference between the merchant and the citizen, between the day-labourer and the citizen, between the landowner and the citizen, between the *living individual* and the *citizen*. The contradiction in which the religious man finds himself with the political man is the same contradiction in which the *bourgeois* finds himself with the *citoyen*, and the member of civil society with his *political lion's skin*.

This secular conflict, to which the Jewish question ultimately reduces itself, [is] the relation between the political state and its preconditions, whether these are [*a*] material elements, such as private property, etc., or [*b*] spiritual elements, such as culture or religion, [and] the conflict between the general interest and the *private interest*, [i.e.] the schism between the *political state* and *civil society*. . . . But the completion of the idealism of the state was at the same time the completion of the materialism of civil society.

Marx 1843: 152–66

1

The Five Symbols of Sikhism

All emancipation is a *reduction* [or restoration] of the human world and relationships to *man himself.*

Marx 1843: 168

The Semiological Method of Study

To begin with a matter of present usage and apparent detail, the custom of wearing long and unshorn hair (*kes*) is among the most cherished and distinctive signs of an individual's membership of the Sikh Panth, and it seems always to have been so. The explicit anti-depilatory injunction was early established as one of the four major taboos (*kuraht*) that are impressed upon the neophyte at the ceremony of initiation into Sikhism, and unshorn hair is one of the five symbols that every Sikh should always wear on his person. Yet there exists hardly any systematic attempt in Sikh studies to explain and interpret the origin and significance of this custom (a noteworthy exception is Kapur Singh 1959: 53–107). As a religious system, mode of thought and code of conduct, Sikhism is anti-ritualistic in its doctrinal content and general tone, so that a study of the few obligatory rites and ceremonies that are associated with it in its institutional or social aspect should be of considerable interest for their own sake. Moreover, if our investigation of the connection between the nature of Sikhism as a whole and its five symbols, including the specific custom of being unshorn, were to be made in a comparative theoretical and empirical spirit and according to rules of method capable of universal application, we may expect that the solution of this particular problem would also illumine certain general problems of the sociology of religion, for example, regarding the nature of religious innovation and its social institutionalization.

I myself am not able to adequately investigate the problem at present since I do not possess the requisite linguistic proficiency to study the original Punjabi and other sources, and without enquiring into them at first hand one cannot proceed satisfactorily. The argument and interpretation presented in this study will be based solely on the information available in English, and for my reliance on second-hand sources that are incomplete and liable to error I render an apology in advance. I shall hope nevertheless that the semiological method or scheme of interpretation that I shall adopt might invoke some interest. For the results achieved, or capable of being achieved, in a line of enquiry depend not only on the evidence examined and its authenticity, but also on the method of analysis and explanation or interpretation followed.

The particular method adopted here, and which may be called the structural method as a variety of semiology, implies that, for a proper theoretical understanding or explanation, the ceremonial custom or rite in question must be viewed from two interrelated aspects. We should attempt to determine (*a*) its theoretical or ideological meaning within a particular cultural or symbolic mode of thought, and (*b*) its effect or social function within a particular code of conduct and social system of groups and categories, in the hope that the two will conjointly explain (*c*) the articulating principles of personality, culture and social organization. The first aspect of thought and belief is chiefly a matter of examining the ceremony or rite as part of a condensed statement, the symbolic expression of certain characteristic cultural facts, ideas and values. In the second aspect of our study we move to the principles of institutionalized behaviour or social action, and seek to relate the rite and the social occasion of its performance to the wider social system of the group or category of persons who recognize the obligation to perform it. In neither case do we consider the particular rite in isolation but bearing in mind the context of the other rites with which it is associated in thought and life, and at either level of analysis our understanding proceeds by seeking to relate the part to its larger whole, the piece to the pattern. Only after these preliminary steps have been accomplished in the context of a particular culture and society, though in the light of general theoretical ideas, may we rightly proceed further to compare the meanings and effects or social functions of similar rites observed in two or more different cultures, or even of the same rite in a single culture at different historical periods.

Combining these two aspects or levels of thought and behaviour, which it is convenient to distinguish for analysis, we may state the central assumption of our procedure in the form that all ceremonies and rites are constitutive, expressive and affirmative in character, that is, they articulate, embody and communicate abstract meanings, facts and values in concrete shape. The obligatory and oft-repeated social performance of a body of rites serves, at the very least, to give definitive expression and form to a people's collective life and thought. It constitutes, presents and affirms to others and to themselves the structural coherence of their particular pattern of culture, thought and social organization as an ordered whole, rather like a language, and contributes to transmit, maintain and develop that pattern from generation to generation. These effects, in short, together constitute, according to our chief theoretical assumption, the *raison d'etre* of ritual behaviour and symbolic thought.

It will be apparent to anyone who has made the attempt that an investigation of the exact meaning, effect and social function of a rite is a complex and difficult task. It is a process like that of ascertaining the grammar and syntax of a language, its structure as against its lexicon, which cannot be done by common sense observation or by simple enquiry from a native speaker or informant. For ritual, if not all custom, is capable by its inner nature of encapsulating several abstract meanings and social references; and moreover these generally do not lie readily accessible at the conscious surface of life but require to be extracted, as it were, from the subconscious or the unconscious, individual and collective. It is therefore specially necessary in this field of study to avoid all easy or ready inferences from intuition or deductive reasoning or commonsense and to adhere to explicitly formulated rules of method.

Symbols, Sects and Initiations

The cultural association of male hair, and specially long hair, with magical or sacred ideas is known from many parts of the world. It is well recognized in general terms to be a symbol of manliness, virility, honour, power, aggression and so on. For example, in very early Europe the Achaeans, who conquered Greece, customarily wore their hair long and wild. The Semitic story of Samson and Delilah as told in the Old Testament well illustrates the virtue of remaining unshorn. We can readily locate many similar examples in classical Hinduism. The Institutes of Manu specify that: 'Even should a man be in wrath, let him never seize another by the hair; when a Brahmin commits an offence for which the members of other castes are liable to death, let his hair be shaved off as sufficient punishment.'

We should, however, be careful to remember that, like all sacred or tabooed objects, long hair can also equally carry the opposite connotation. It can be regarded, especially when unkempt, as signifying something unclean, dangerous or abandoned. We must thus refer, according to the rules of our method, to the actual context and situation in order to determine which of these two elements is predominant in a particular case.

That the precise physical state of the hair is always relevant to its symbolic meaning, but is never by itself the deciding factor, can be made clear from the example of the Chinese pigtail, which superficially resembles the Hindu *shikha* (scalp-lock) in appearance. The Manchus, a foreign dynasty ruling in China, in fact first instituted the pigtail among the Chinese in AD 1644 as a sign of their subjection. It later became accepted as a characteristic Mandarin custom, even as a sign of honour. In the mid-nineteenth century the Taiping rebellion and in the early twentieth century Sun Yat-sen's movement and others sought to dispose of it, remembering its original significance. The Taipings did so by wearing all their head hair long and so became known as the 'long-haired rebels', whereas the twentieth century revolutionaries proceeded to cut all their hair short, literally throwing the pigtail away. The complete contrast between these two outcomes of a single impulse is not without interest for our study.

In Sikhism the injunction to remain unshorn is expressly associated with the ceremony of initiation, and it is in that context that we must primarily explain it. Now every initiation rite evidently possesses the nature or effect of an investiture or conferment, since through it some new status or role with its consequent rights and obligations is conferred symbolically upon the neophyte and he or she enters upon a new mode of existence. But every initiation rite necessarily also contains a much less obvious element, namely that of renunciation or divestiture, whereby the neophyte symbolically discards or has taken away from him attributes of his old status and mode of existence. One must ritually first abandon, in other words, the previous course or phase of social existence in order to properly enter the new. Admittedly, the positive element of investiture or conferment generally predominates in initiation rites, but the element of renunciation or divestiture is always present to some degree. This negative element may even be uppermost in certain cases, for example, in initiation to monkhood or the monastery.

I now want to draw attention to a class of initiation rites of this latter kind that were widely prevalent in the Punjab and elsewhere at the time that Sikhism took its origin. These were rites of renunciation (*sannyas*) through which an individual obtained entrance to one or other of the medieval mendicant orders (Sannyasis, etc.). It is my contention that an examination of this class of rites with the details of the Sikh initiation rite borne in mind, shows a remarkable relation of structural inversion to exist between the two. I want to suggest that, in terms of the symbolic language and ritual idiom of the times, at least one cultural meaning and effect of the Sikh initiation rite was that it stood as the antithesis or the antonym of the rites of Hindu renunciation.

A *sannyasi* is a person who, having passed through the first three statuses (*ashramas*) of Brahmanical Hinduism, renounces the world and is cared for by others. It may perhaps be that the Sannyasi religious orders were older than the Brahmanical institution of *sannyas*, the fourth and last stage of life. At any rate, the orders seem to have been open to entry by the individual person of almost any physical age. The Sannyasi orders had decayed significantly during the Buddhist period and then split into sub-orders often with heterodox creeds. They were reformed by

Shankaracharya, whose four chief disciples instituted four *maths* (orders) that later developed into numerous *padas* (sub-orders). Each sub-order was said to be constituted of two sections, one celibate and mendicant, the other not. All individual Sannyasis were further graded according to four degrees of increasing sanctity (Kavichar, Bahodak, Hans, Paramhans).

The Sannyasi initiation rite was and continues to be essentially as follows (the ethnography presented in the succeeding paragraphs is derived from Rose, 1911 and 1914, whose compilation is based on the census reports for the Punjab, 1883 and 1892). The candidate intending to attain renunciation must first go on a pilgrimage to find a guru, who should be a Brahmin; and then the latter on his part satisfies himself as to his fitness and proceeds to initiate him. The neophyte commences with the *shraddha* (obsequies) to his ancestors to fulfil his obligations to them. He next performs the sacrificial *baji hawan* and gives away whatever he possesses, severing all connection with the social world. His beard, moustaches and head are entirely shaved (*mundan*), retaining only the scalp–lock (*shikha*), and the sacred thread is put aside. He then performs the *atma–shraddha* or his own death rites. (An initiated Sannyasi is thus counted as socially deceased, and when he dies later is not cremated but buried in a sitting posture without further ceremony.) The scalp–lock is now cut off and the neophyte enters the river or other water with it and the sacred thread in hand and throws them both away, resolving, 'I am no one's, and no one is mine.' On emerging from the water he starts naked for the north but the guru stops him and gives him a loin–cloth (*kopin*), staff (*danda*), and water vessel (*jalpatra*) kept out of the neophyte's personal property. Finally, the guru gives him the *mantra* (spiritual formula) in secret and admits him to a particular *math* (order), *sampradai*, etc. (Rose 1914: 358).

The initiation rite of the Jogi order, which was also widespread in medieval Punjab, is very similar. According to the *Punjab census report*, 1912, 'Jogi' is a corruption of *yogi*, a term applied originally to *sannyasis* well advanced in the practice of *yogabhyas*.

The Jogis are really a branch of Sannyasis, the order having been founded by Guru Machhandar (Matsyendra) Nath and Gorakh Nath Sannyasis, who were devoted to the practice of Yoga and possessed great supernatural power. The followers of Guru Gorakh Nath are absorbed more in the Yoga practices than in the study of the Vedas and other religious literature, but between a real good Jogi and a *yogi* Sannyasi there is not much difference, except perhaps that the former wears the *mudra* (rings) in his ears. The Jogis worship Bhairon, the most fearful form of Shiva (Hari Kishen Kaul, *Punjab census report*, 1912, quoted in Rose 1914: 361). [Their main subdivisions are stated to be the Darshani or Kanphatta (split-eared), known as Naths, who wear the *mudra* (ear-rings); and the Aughar, who do not.]

In Jogi initiation the neophyte (*chela*) is first made to fast completely for two or three days. A knife is then driven into the earth, and the candidate vows by it not to (*a*) engage in trade, (*b*) take employment, (*c*) keep dangerous weapons, (*d*) become angry when abused, and (*e*) marry. He must also scrupulously protect his ears, for 'a Jogi whose ears were cut used to be buried alive, but is now only excommunicated'. The neophyte's scalp-lock is removed by the guru and he is shaved by a barber; his sacred thread is also removed. He bathes and is smeared with ashes, then given ochre clothes to wear, including the *kafni* (shroud). The *guru-mantra* is communicated secretly, and the candidate is now a probationer (*aughar*). After several months' probation his ears are pierced and ear-rings inserted by the guru or an adept, who is entitled to an offering of one and a quarter rupee. 'The *chela*, hitherto an *aughar*, now becomes a *nath*, certain set phrases (not *mantras*) being recited' (Rose 1911: 400). The Jogis hold the element earth and everything made of it in great respect. 'The earthen carpet, the earthen pitcher, the earthen pillow and the earthen roof', is a saying that describes their life. Like the Sannyasis, Jogis are customarily buried in the earth and not cremated.

According to an account of the Ratan Nath Jogis, the intending candidate is proffered a razor and scissors by the guru to deter him from entering the order. If he perseveres the guru cuts off a tuft of his hair and he is shaved by a barber. He is made to bathe, smeared with ashes and then given a *kafni* (shroud), a *lingoti* (loin-cloth) and a cap to wear. 'The ashes and *kafni* clearly signify his death to the world.' After six months' probation his ears are pierced and earthen ear-rings inserted in them (Rose 1911: 401n).

After his initiation, a Jogi may either remain a celibate and ascetic mendicant (called Nanga, Naga, Nadi, Nihang or Kanphatta), living on alms; or he may relapse and become a secular Jogi (called Bindi-Nagi, Sanyogi, Gharbari or Grihisti), holding property and having a spouse. A Jogi usually joins one or other of the various *panths* or 'doors' (sub-orders), whose traditional number was stated to be twelve.

I mention finally the initiation rite of the Dadupanthi order, stated to have been founded by Dadu, a Gaur Brahmin who died in AD 1703. In this rite the guru in the presence of all the *sadhus* shaves off the neophyte's scalp–lock and covers his head with a skull cap (*kapali*) like the one that Dadu wore. The latter dons ochre clothes and is taught the *guru-mantra*, 'which he must not reveal' (Rose 1911: 215f). The Dadupanthi rite concludes with the distribution of sweets. It is said that other accounts make Dadu contemporary with Dara Shikoh, and still others with Guru Gobind Singh. The *Gur-bilas* gives an interesting story about Guru Gobind Singh's meeting with a Dadupanthi (see Indubhusan Banerjee 1962: 94f).

Now, for any exercise in historical sociology, the 'book view', giving priority to texts—classical, medieval or modern, versus the 'field view', relying on observation of institutions and behaviour, both require to be confronted. In this case, the monograph by the Orientalist Jan Gonda on *diksha*, the Hindu ordination or consecration which 'answers to baptism as well as "initiation" in other religions', does not tell any different story (1965: 315–462). The same elements and relations appear as in the census reports, which I have assumed to be based on observation, and together they deliver the same message. I have formed the impression that Gonda, who follows the historico–philological method, would have done better service to mutually compare (*a*) the Veda student's *upanayana* (a Brahmin's coming of age sacred-thread ceremony, although *diksha* is not found in the Rig Veda), (*b*) initiation into *sannyas*, 'complete casting off', or into monkhood, e.g. the Buddhist order or *sangha*, and (*c*) the consecration of the king. But one must make do with what one has, so I simply abstract below what Gonda has said, giving equal importance to the classical or ancient texts, the Upanishads, as well as to medieval texts of the sectarian triad, Shaivism, Vaishṇavism and Tantrism.

Now [in classical times], a group of later upanisads [ed. F.O. Schrader] which deal with these wandering almsmen and their condition of life, interestingly enough, prescribes for them an initiatory ritual. Although the details of this ritual vary in different texts, its significance is clear: it is a ceremonial rejection not only of home–life, but of the whole system of Vedic socio–religious practices as observed by the householders who regulated their lives in accordance with the dharma established by the brahmans. . . . It is indifference to all worldly objects (*vairagya*) and the realization of the oneness of individual and brahman [identity of the individual soul and the world soul] which make a man a true samnayasin. But this condition of life, like the various stages of existence in the world, can only be entered upon by passing through an "initiation".

In illustration attention may be invoked to the so–called Samnyasa–Upanisad which was at the time translated by Paul Deussen.

. . . The man who desires to pass beyond the normal stages of life (*asrama*) should sanctify himself with Vedic mantras. He should, for the last time, perform, in the forest, a sacrifice in honour of the Fathers. . . . After having by this rite taken leave of the world, he places his three sacrificial fires in himself.

. . . On the difference between the *grhastha* who is an *asramin* and the [initiated] *samnyasin* . . . [who] puts on the brown-red ascetic's habit, removes the hair from arm-pit and abdomen and keeps his arm raised. Two usual practices recur: shaving and another dress. . . . According to the Paramahamsa-Upanisad (2) an ascetic of the highest order (*paramahamsa*) should live without staff, clothes, tuft of hair and sacred cord. The true samnyasa however consists in the realization of the oneness of the individual "soul" and the soul of the All.

As will be seen further certain Tantrist initiation rites, known from later sources, involved the performance of one's own funeral rites. It may also be recalled that the brahmana texts dealing with the classical soma sacrifice contain unmistakable allusions to the self-sacrifice and death of the consecrated *yajamana* or of other consecrated persons . . . "verily taking the self as the gift they go to the world of heaven".

Visnuism felt the need of initiation in the life of devotion very strongly. . . . The first initiation into Visnuism is called the Pancasamskara ceremony. . . The most important element of this ceremony is the recitation of the mulamantra and the "double mantra", which are held to be most sacred and very important by all Sri Vaisnavas, in the right ear of the initiate in such a low tone that it can be heard only by himself.

The compiler of the Varaha-Purana . . . deals with the diksa in case the aspirant is not a brahman. The ksatriya who wishes to undertake the ceremony should give up all his weapons. . . . A vaisya should likewise renounce his customary way of life and devote himself to God's ritual work. . . . A sudra who undergoes the Visnudiksa is freed from all sins. . . . All persons including the sudras could repeat the Ramamantra.

Diksa is a very important ceremony in all schools of Sivaism. When, in the course of transmigration, the soul has reached a considerable state of purity so as to strive after omniscience—which leads to emancipation—Lord Siva, extending his grace, instructs the soul, manifesting himself as an internal light or as a guru. . . . It is therefore the faith of the "non-Vedic" (*avaidika*) Sivaites that he who has not received the *saivi diksa* does not attain moksa. . . . Then the soul begins to abhor worldly life and to take interest in the attainment of emancipation. It finds a competent guru, who gives diksa which ultimately disentangles it from the *pasas*, destroying the corruptive power of mala. Diksa is thus regarded as the most essential condition of emancipation. It removes the soul's *pasutva* ("animality") and restores it to its pristine *sivatva* or divine nature (Gonda 1965: 377ff, 384, 398, 401f,.409, 418f, 429, 433).

The Sikh Initiation and its Five Symbols

In my view, there can be little doubt in the light of the evidence that the anti-depilatory taboo (*kuraht*) of the Sikh initiation rite is to be understood as a specific inversion in structural terms of the custom of total depilation enjoined by the Jogi, Sannyasi, etc. initiations. The element of symbolic inversion, as I see it, is in fact much more pervasive, but it has been entirely overlooked before owing to the prevalence, among students of religions, of the scholarly method of endlessly adducing parallels and similarities to the neglect of significant relations of reflection, contrast, counterpoint, inversion and opposition (this neglect is apparent, for example, in van Gennep 1960: 97, and the same method is followed by Kapur Singh 1959: 85ff). In contradistinction to the Jogi and Sannyasi ritual of nakedness or smearing with ashes, the Sikh neophyte is made to come tidily clothed to the ceremony. The ear-rings affected by the Jogis are specifically forbidden to him (Teja Singh 1938: 113). Instead of requiring the Sannyasi's resolve, 'I am no one's, and no one is mine', the Sikh rite, emphasizing a new birth, requires the neophyte to reply in answer to questions that his father is Guru Gobind Singh and his mother Mata Sahib Kaur, and that he was born in Kesgarh and lives in Anandpur. Even more significantly perhaps the initiator, instead of being the individual guru, is a collective group, the 'five loved ones/the five lovers', composed of any five good Sikhs. Instead of the *guru-mantra* being communicated privately and secretly to the neophyte, as with the Sannyasis, Jogis and Dadupanthis, the Sikh gurus' word is spoken loud and clear in public congregation by the initiators. Finally, in contrast to the Jogi vow never to touch weapons, the Sikh neophyte is invested ritually with the *kirpan* (sword) as one of the five *k's* which he must always wear thereafter.

I think one may safely say that the Sikh initiation rite contains a marked theme of inversion in relation to the rites of social renunciation established by the medieval mendicant orders that preceded Sikhism. Like them, Sikhism was instituted as a religious brotherhood open to all who sought salvation, but its spiritual and social aims were in direct contrast to what theirs had been. Whereas they had sought to achieve emancipation and deliverance through individual renunciation and what amounted to social death, as their rites signify, the new Sikh community was called to affirm the normal social world as itself the battleground of freedom. The three traditional functions of the guru, as the link in a chain of sacred, transcendental or immanent beginnings, a mediator who can bring God and his pupil together and as a medium through whom God is willing to reveal himself, are all here aspects of the collectivity rather than the individual. That is why the Sikh initiation rite makes the positive theme of investiture prevail wholly over the negative theme of divestiture, and taking certain widely established customs of Hindu renunciation, emphatically inverts them. The meaning of being unshorn, in particular, is thus constituted in this analysis by the 'negation of the negation': it signifies the permanent renunciation of renunciation as a principle.

This semiotic hypothesis, however, is not yet complete; it requires a further consideration regarding all the five *k's*. We have so far concentrated our attention on the initiation rite itself and attempted to understand the meaning of *kes* in that context, but the five symbols of Sikhism are worn for life. Now, following initiation, the Sannyasi custom is to either wear all their hair or shave it all. The *jatadhari* Jogis follow the former course—though among all Jogis the signification of renunciation seems to be borne primarily by their pierced ears and ear-rings. The important order of Bairagis also keep long hair, for example, whereas the Uttradhi Dadupanthis shave the head, beard and moustache (Rose 1911: 36, 216). The Rasul Shahis, a Muslim order founded in the eighteenth century, also shave completely the head, moustaches and eyebrows (Rose 1914: 324). In all such instances, wherever long hair is worn, it is worn as matted hair (*jata*), frequently dressed in ashes. According to Sikh custom, on the other hand, unshorn hair (*kes*) is invariably associated with the comb (*kanga*, the second of the five *k*'s), which performs the function of constraining the hair and imparting an orderly arrangement to it. This meaning and effect are made even clearer by the custom of the Sikh turban, worn enclosing both the *kes* and the *kanga*. The *kes* and the *kanga* thus form a unitary pair of symbols, each evoking the meaning of the other, and their mutual association explains the full meaning of *kes* as distinct from *jata*. The *jata*, like the shaven head and pierced ears, symbolizes the renunciation of civil society or citizenship; the *kes* and *kanga* together symbolize its orderly assumption.

The *kirpan* (sword) and the *kara* (steel bracelet) similarly constitute another pair of symbols, neither of which can be properly understood in isolation. Without going into the evidence, I merely state that in my view the bracelet imparts the same orderly control over the sword that the comb does over the hair. The medieval ascetic order of the Kara Lingis indeed wore on a chain a similar ring over the naked penis. The *kirpan*, in its conjoint meaning with the *kara*, is a sword ritually constrained and thus made into the mark of every citizen's honour, not only of the soldier's vocation. Finally, the *kachh*, a tailored loin and thigh garment, the last of the five *k*'s is also to be understood as an agent of constraint, like the comb and the bracelet, though the subject of its control is not overtly stated. This unstated term, I think, can only be the uncircumcised male or female member.

The *kachh* constitutes a unitary pair of meanings with it, signifying human reserve in commitment to the procreative world as against renouncing it altogether.

In case it might be objected that I am merely profaning the mystery in advancing the last hypothesis, I hasten to quote Guru Gobind Singh himself as reported on the subject:

> Ajmer Chand inquired what the marks of the Guru's Sikhs were, that is, how they could be recognized. The Guru replied, "My Sikhs shall be in their natural form, that is, without the loss of their hair or foreskin, in opposition to ordinances of the Hindus and the Muhammadans" (Macauliffe 1909: V, 99).

We can now formulate the proposition that the primary meaning of the five symbols, when they are taken together, lies in the ritual conjunction of two opposed forces or aspects. The unshorn hair, the sword and the implicit uncircumcised male organ express the first aspect. They are assertive of forceful human potentialities that are of themselves amoral, even dangerous, powers. The comb, the steel bracelet and the loin and thigh breeches express the second aspect, that of constraint and moral discrimination. The combination of these two aspects is elaborated in the form of three pairs of polar opposites (*kanga/kes* :: *kara/kirpan* :: *kachh/*the uncircumcised state), thus generating, with one term left unstated, the five Sikh symbols. The aspect of assertion and the aspect of constraint combine to produce what we may call for want of a better word the spirit of *affirmation* in history and society, characteristic of Sikhism as an example of modern non-dualism.

Sikhism and Hinduism

So much then for the structural explanation of cultural meanings, the logic of symbols. We must now turn, although very briefly and simply, to the second level of analysis required by our method, and consider the wider social context of Sikhism's origin and growth. I do not here give all the evidence or make every qualification but state the problem in broad and general terms as follows. The Hindu system of social relations called caste, using that term to include *varna* as well as *jati*, is in fact only the half of Hinduism. The whole Hindu *dharma* is better described by the term *varnashramdharma*, that is, caste as well as the institution of the four stages, statuses or roles of individual life (*ashramas*). If sociologists have hitherto concentrated on the institution of caste to the exclusion of the latter institution, which is left to be celebrated by Indologists, I cannot claim to understand their reasons. For the social system of caste was always surrounded in India by a penumbral region, as it were, of non-caste, where flourished the renunciatory religious orders whose principles abrogated those of caste, lineage and birth; and the fourth *ashrama* (*sannyas*) constituted a door through which the individual was recommended to pass from the world of caste to that of its denial. The mutual relation of the two worlds, and I have no doubt that it was mutual, is of the greatest significance to a full understanding of either of them. Caste, and particularly the position of Brahmins, was stated by Max Weber to be 'the fundamental institution of Hinduism' (1948: 396). This purely one-sided view is especially curious in the German sociologist since he was the first to make use in 1916 of the 'partly excellent scientific Census Reports', as he says, which also form the basis of Rose (1911 and 1914) and of the present study.

The hierarchical system of local caste groups, with membership predetermined by birth, on the one hand, and the wholly contrasted system of voluntaristic cult associations of orders and sects, spread over wide areas, on the other hand, always cut across one another, forming the essential warp and woof of Hinduism. The third structural feature, territorial kingship, necessary to uphold *varnashramdharma*, possessed its own relations with the two contrasting worlds of the Brahmin and the *sannyasi*. Thus the total medieval Hindu world, including its political institution, effectively produced and rested upon a tripartite division and a system of interrelations among the three separate worlds symbolized by the king, the Brahmin and the *sannyasi*. The domains, statuses or roles of (*a*) the rulers, the world of *rajya*, (*b*) the caste system, *varna* or *grihasta*, and (*c*) the orders of renunciation or *sannyas*, formed the three sides of the medieval triangle. Moreover, the same principles or underlying structure can perhaps be seen in the Islamic culture of the period in the division and interrelations among the three respective spheres of *hukumat* (the state power), *shari'at* (the social order) and *tariqat* or *haqiqat* (the Sufi sect, taken as a way of salvation). These are all sociological problems for further investigation, as in the next chapter, when broken down into suitable units for study.

An order, let us say, like the Aghor-panthi Jogis, who appear to have smeared themselves with excrement, drunk out of a human skull and occasionally dug up the body of a newly-buried child to eat it, 'thus carrying out the principle that nothing is common or unclean to its extreme logical conclusion', evidently constituted the truly living shadow of Hindu caste orthodoxy. The theme of antinomian protest could hardly be carried further—unless it was by the Bam-margis who added sexual promiscuity to the list! Yet it could be reliably said of other Jogi sub-orders that 'in the Simla hills the Jogis were originally mendicants, but have now become householders', and that the secular Jogis, called Sanyogis, 'in parts of the Punjab form a true caste' (Rose 1911: 399n, 404, 409). We can resolve this seeming contradiction, I think, only if we regard both these Jogi conditions as forming the different stages or phases of a single cycle of development. According to this view, one should say that any particular order or sub-order that once renounced caste with all its social rights and duties, and walked out into the ascetic wilderness through the front door of *sannyas*, could later become disheartened or lose the point of its protest, and even end by seeking to re-enter the house of caste through the back door.

Of course, I would say that as a particular order or section fell back, so to speak, from the frontier of asceticism and abandoned its non-procreative, propertyless and occupationless existence, its function within the total system of *varnashramdharma* would be fulfilled by some other order or section, since the ascetic or protestant impulse itself remains a constant feature in history. During its ascetic period, an order or sub-order may occupy one or the other of two positions, or pass through both of them successively. It may either adopt a theory and practice completely opposed to those of caste, like the Aghor-panthis and Bam-margis, and be for that reason regarded as heterodox and esoteric; or it might remain within the pale and link itself to the caste system through the normal sectarian affiliations of caste people, who adopt a guru but remain householders. An 'heterodox', left-handed or antinomian sect, we should say, is one opposed to caste as its living shadow; an 'orthodox' or right–handed sect is complementary to the caste system, its other half within Hinduism.

I would not say that all historically-known orders of renunciation in fact passed through these various stages of development, but I maintain that we must construct some such analytical scheme of their typical life–history with reference primarily to their origin, function and direction of movement in relation to the caste system, or more accurately ***varnashramdharma***. For that would enable us to classify the vast number of known sectarian orders and sub-orders into a limited number of sociological types, and obviate many difficulties in our study of them. In particular, until we can fully understand the developmental cycle of medieval mendicant orders we cannot place the *political* phenomena of the 'fighting Jogis' of the sixteenth and seventeenth centuries, the long contemporary militant struggles of the Islamic Roshaniyya sect (founded by Bayazid, Pir-i Roshan, b. Jullundur, 1525), the Satnami revolt of 1675, or the plunder of Dacca in 1763 by Sannyasis, etc., in their proper structural perspective. The analytical paradigm proposed must account under one and the same theory for cases, types or phases of political quietism as well as of political activism and conflict.

As a social movement early Sikhism, I have no doubt, possessed many features in common with other religious brotherhoods of a certain type. If Sikhism as a whole nevertheless broke free from the convoluted cycle of caste versus non-caste that overtook other protestant or antinomian brotherhoods, to what cause or causes did it owe its freedom? It is true that Sikhism, as others have noted earlier, barred the door of asceticism and so did not lose itself in the esoteric wilderness, but we have also to explain why it did not duly return, as so many others did, to the citadel of caste. The new departure of Sikhism, in my structural interpretation, was that it set out to annihilate the categorical partitions, intellectual and social, of the medieval world. It rejected the opposition of the common citizen or householder versus the renouncer, and of the ruler versus these two, refusing to acknowledge them as separate and distinct modes of existence.

It acknowledged the powers of the three spheres of *rajya, sannyas,* and *grihasta*, but sought to invest their virtues conjointly in a single body of faith and conduct, religion-in-society-and-history, inserted by grace and effort as mediation between heaven and the world, or the *atma* and *Paramatma*, the individual and the All, as the modern Indian form of non-dualism of self, the world and the other.

The social function and effect of the Sikh initiation rite is, I think, precisely this: to affirm the characteristic rights and responsibilities of the three different spheres as equally valid and to invest them as an undivided unit in the neophyte. The new male Sikh, therefore, takes no Jogi vow to renounce his procreative power and never marry; instead he dons the manly *kachh* of continence. Instead of vowing like the Jogi never to touch weapons or take other employment or engage in trade, every social occupation is henceforth open to him, including that of soldiering, householdership or political command, and save only that of renouncing productive labour and taking alms. The same or similar rules apply to the female neophyte, but the matter has been insufficiently reported upon. The single key of the 'renunciation of renunciation' was thus charged to unlock all dividing doors in the mansion of medievalism; whether it succeeded in doing so and to what extent is, however, an open question.

The semiological or structural method of analysis and interpretation, of which I have attempted to provide an example, shows that we can establish a definite connection between the five symbols of Sikhism and its whole nature as a religion. If my previous pairing of symbols and the assumption of an unstated term be accepted, then the five symbols of Sikhism may be said to signify, in their three respective pairs, the virtues and the roles of *sannyas yoga* (*kes* and *kanga*), *grihasta yoga* (*kachh* and the uncircumcised state) and *rajya yoga* (*kirpan* and *kara*), whether or not these spheres are further taken to correspond to religion as a way of salvation, the life of civil society and the state. As the authenticating sign and seal of Sikhism, the five *k*'s together affirm the unity of man's estate as being all of a piece: this we may take to be the final meaning and effect of remaining forever unshorn in the world. Our analysis would also lead to the conclusion that the total human emancipation of religious man, and not merely any ideal of a synthesis or reconciliation of Hinduism and Islam, was the faith and endeavour of Sikhism from its inception. The institutionalization of that faith and endeavour surely marked the opening of the modern period of history in the Punjab.

2

The Elementary Structure of Medievalism

When religion was individualized and got detached from the state and the family the propensity for mystic cults rose. . . . They promised the transfiguration of their devotees, deliverance and a happy after-life.

Whereas the Vedic Aryan tradition sanctions and extols the householder's life, the *grhastha* found his opposite in the *parivrajaka*, the wandering religious mendicant, who rejecting the dharmabooks of the brahmans, "giving up children, brothers, relatives, the sacred cord and the brahmanical tuft of hair, Vedic-study as well as sacrifices" (Aruni-Upanisad 1) went "from home into homelessness"—which is considered to be a state of purity and freedom leading to salvation. . . . In the centuries in which the Vedic culture faded into [medieval] Hinduism [after Buddhism] . . . the wandering ascetics and their hermitages contributed much to the cultural colonization of the Indian peninsula. In customs and way of life the ascetics of "orthodox" brahmanic origin did not, as far as may be judged from appearances, stand out considerably from the heterodox recluses and wanderers, however much diversity of opinion there might, among these ascetics, have existed with regard to view of life, soteriology and ascetic practices.

. . . There is from various sources some evidence that they sometimes organized well-knit bodies, observing rites, ceremonies and rules of discipline, professing a particular faith or doctrine, and recognizing one of their members as their leader.

Gonda 1965: 377, 383, 393f

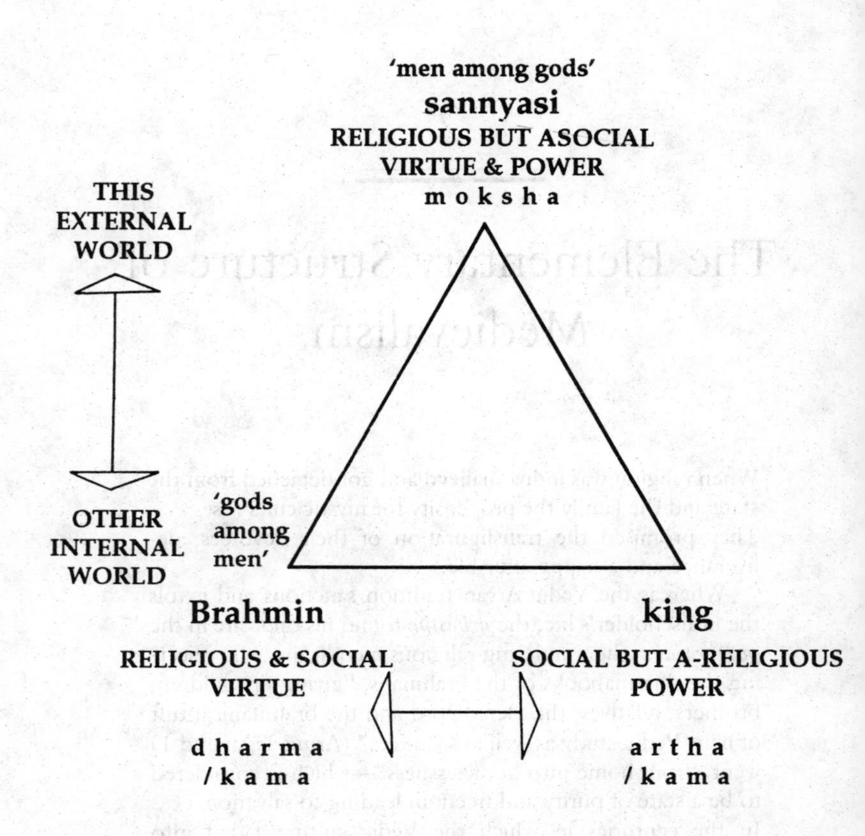

FIG. 1 The Hindu Culture of India

India today is, by common or general consent, a multi-religious nation, a modern pluralist society and a federal secular state; and the sociology of India would surely benefit from a project to view the phenomenon in its two dimensions of synchrony as well as diachrony. A preliminary investigation in that direction, such as the one offered here, need only suppose that the Indian modernity is a transformation of medievalism, howsoever the latter is defined in terms of periods of history, and without denying the permanent Indian inheritance. Personally, and also as a nationalist, I am persuaded that, although the Hindu culture and the Muslim culture always possessed their respective distinctive features and internal systems of coherence, they also possess the same underlying structure of medievalism, even if I cannot precisely date it. If we view the period of medievalism in synchrony, i.e. as a system of simultaneity, the Hindu symbols of the Brahmin, the king and the *sannyasi*, on the one hand, and of Muslim *shari'at*, *tariqat* and *hukumat*, on the other hand, will form our two manifest objects of study for the initial situation of the Indian modernity.

The Hindu Culture of Medieval India

The starting point for our investigation of the elementary triad, the Brahmin, the king and the *sannyasi*, will be two social science papers by Dumont, combining sociology, social anthropology and Indology, based upon his lectures given at Oxford (1958) and in London (1961), one on the renunciation of the world by the *sannyasi* (medieval) and the other on the conception of kingship in India (ancient), in both of which the Brahmin is the implicit other term of the opposition or relation. These two papers of 1960 and 1962 were written some years before the completion of Dumont's now standard work on the Hindu caste system published under the Latin title, *Homo hierarchicus*, first in French (1966) and then in English (1970), and appear to me to go beyond his unduly narrow view of hierarchy, purity and impurity or the encompassing and the encompassed, as the sole logic underlying Indian history and culture.

The Brahmin and the sannyasi

Dumont sees here that Hinduism is the religion of caste *and* of its renunciation, i.e. the group religion of society and its three worldly ends of life, religious duty (*dharma*), power and wealth (*artha*), and pleasure (*kama*), as opposed to the 'disciplines of salvation', the fourth end of liberation (*moksha*) from the flow of existences, transmigration or the law of rebirth for actions, *karma*. It is true that the literature of *dharma* sometimes 'superimposes' the supreme end of liberation, in this sense, upon the other three ends of life, but the bringing together of all four *purusharthas* remains a mere aggregation in theory and practice. It merely masks the 'heterogeneity' between the three legitimate and necessary worldly ends and the negation of the world which, although it is optional, is fatal to the other three once it is adopted by the individual, except perhaps in the way of *bhakti* (1960: 41, 43, 45, 61; 1962: 67n).

The analogy with the hierarchy of *varna* [the caste system] is apparent: *dharma* corresponds to the Brahman or priest, *artha* to the king or kshatriya, the temporal power, and *kama* to the others.

Renunciation is often individually represented as the fourth and last stage or status (*ashrama*) in the life of a Brahmin, who is thus successively a novice, householder, hermit and finally a *sannyasi*.

The Brahmans, as priests superior to all other men, are comfortably enough settled in the world. On the other hand, it is well known that classically whoever seeks liberation must leave the world and adopt an entirely different mode of life. This is an institution, *samnyasa*, renunciation, in fact a social state apart from society proper. The ultramundane tendency does not play only in the minds of men in the world, it is present, incarnate in the emaciated figure of the renouncer, the *samnyasin*, with his begging bowl, his staff and orange [ochre] dress. We may imagine the reaction to this creature of the typical Brahman whom we may think of as he is still represented in a carving on the north gate at Sanchi (Vesantara Jataka), a round bellied figure, expressing an inimitable blend of arrogance and avidity.

It is also true that what the *sannyasi* renounces is the social world (*samsara*) and not the (non-human) material universe of nature (*jagat*), as remarked by J.F. Staal, but Dumont makes nothing further of it. He infers that the *sannyasi* is the 'safety valve' for the Brahmanic order, even though Tantrism as distinct from Vaishnavism and Shaivism, with which it forms the great sectarian triad of Hinduism, may claim through its evident reversal of values to transcend the opposition of *yoga*, discipline, and *bhoga*, enjoyment. 'By means of this compromise the Brahman rules over the world in peace, as a rather monotonous immanence' (1960: 44f, 52f).

What is in the West called the relation or the opposition of society and the individual has its counterpart, therefore, in the Indian dualism of the man-in-the-world and the renouncer, the *sannyasi*, and 'the secret of Hinduism may be found in the dialogue between the renouncer and the man-in-the-world'. It has two faces or facets, 'one for the man-in-the-world, who is not an individual, the other for the renouncer, who is an individual-outside-the-world'.

> The renouncer leaves the world behind in order to consecrate himself to his own liberation. He submits himself to his chosen master [*guru*], or he may even enter a monastic community, but essentially he depends upon no one but himself, he is alone. In leaving the world he finds himself invested with an individuality which he apparently finds uncomfortable since all his efforts tend to its extinction or its transcendence (1960: 37, 46f, 62).

The *sannyasi* or the renouncer does not deny the religion of the man-in-the-world: they are related or opposed essentially as the individual to the collective.

> Here then lies the possibility of aggregation: the discipline of the renouncer by its very tolerance of worldly religion becomes additional to it. An individual religion based upon choice is added on to the religion of the group.
>
> It seems *a priori* that there are two ways open to the man who leaves the world and finds himself endowed, as a result of his renunciation, with an individuality. [*a*] He can assume this individuality in order to end it by liberation, this is the way of Buddha, who only maintains the liberty of man for this end (it being granted that out of compassion one may delay this end). Or [*b*] he can accept his individuality to control and order it, such would seem to be the way corresponding to *samkhya* dualism and the monotheism of *bhakti* (1960: 46, 51n).

Historically, the path of devotion appears early in the middle ages of India, 'since after one Upanishad the Bhagavad Gita reveals it and has remained as its Bible'.

It expounds in [ascending] order three paths to union or disciplines of salvation: [*a*] that of acts, [*b*] that of knowledge and [*c*] that of devotion. The first two correspond respectively to life in the world and to renunciation, but they are modified and even transmuted by the intervention of the third . . . for it is the discovery of this alone which makes possible the attainment of salvation through action.

The central point is that, thanks to love, renunciation is transcended by being internalized; in order to escape the [*karma*] determinism of actions, inactivity is no longer necessary, detachment and disinterestedness are sufficient; one can leave the world from within, and God himself is not bound by his acts, for he only acts out of love . . . [and] by loving submission, by identifying themselves unreservedly with the Lord, everybody can become free individuals.

But whereas the *bhakti* of the Bhagavad Gita is speculative and intellectual, medieval Hindu *bhakti* proper, met with later in the Tamil hymns and in the Bhagavata Purana in Sanskrit, 'is on the contrary highly emotional'. Dumont considers it therefore to be like possession, 'a functional feature of folk religion', and I would say therefore it is also individual in its character.

But all the prestige and fecundity of renunciation end up by offering to the man-in-the-world a choice of religions for the individual. At the end of the movement—an end achieved very early, with *bhakti*—the renouncer is in fact absorbed, whether he invents a religion of love open to all, whether he becomes the spiritual head of worldly people, rich or poor, or whether he remains a Brahman while becoming a sanyasi, as with Ramanuja. By this point, if to every one the path of spiritual adventure is as open as ever, socially the circle is closed (1960: 57ff).

The fact of a religion of individual choice being 'superimposed' upon group religion is similar to what one finds in classical European antiquity, e.g. as when the Greeks went to Eleusis for initiation. But we must ask Dumont for a final word about the sociology of Hinduism as the relation between the caste system and its apparent negation, the sect, the typical Indian relation to which I want to draw attention also among the Muslims under the regime of medievalism.

Practically all the sects have been founded by sanyasis and the greater part include, apart from the worldly adherents, a sanyasi order which constitutes the nucleus of the sect. The link between both sides is provided by the ancient institution of the spiritual master or *guru*. Instead of the renouncer alone having his *guru*, he serves in his turn as *guru* to whoever wishes. The institution is thus most remarkably enlarged or democratized. A majority of Indian heads of families, of all castes—even Muslims—have chosen a *guru* who has initiated them while whispering a *mantra* in their ears, and who, in principle, visits them once a year.

A second feature of the sect is that, unlike orthodox Brahmanism, it is not essentially syncretic but holds to one doctrine, the principle of its unity. Often, in fact it is monotheist in the true sense, that is it denies other gods, the gods of others, and is not content with pushing them into the background. Also, as is well known, the sect, whatever its dominant inspiration, transcends caste and is open, at least in principle, to all, as is appropriate in a creation of the renouncer.

Let us compare here orthodox Brahmanism and the sect. On the one hand there is a multiplicity of gods (or a speculative pantheism), syncretism and considerable tolerance in the domain of belief, tolerance that is of the *object* of religion, which contrasts with the strict exclusivism as regards the *subjects* of religion, the people who can be admitted amongst the faithful. The sect, on the other hand, is inclusive as regards the subjects, the faithful, strict and exclusive as regards the god and belief, the object of religion. Indian multiplicity has then its limits, it only bears upon one of the two poles that make a religion. Abstractly considered sect and Brahmanism appear as two variants of a disposition balancing multiplicity and unity, inclusion and exclusion.

Dumont concludes that the real and true history of Hinduism lies in the sannyasic developments, on the one hand, and in their 'aggregation' to worldly religion, on the other hand. 'Where we condemn and exclude, India hierarchizes and includes': he forgets here the role of separation as the other half of hierarchy. Thus, medieval Hinduism is completely built up in the period which saw the decline of the great heresies, like Buddhism, by 'the progressive integration or, as I prefer to call it, aggregation by orthodox tendencies of elements introduced by the heterodox' (1960: 36, 42n, 47n, 59f).

My conclusion from this restructured presentation of Dumont's argument and evidence is that it is not inconsistent with what I have said in the first chapter of the relation of Hinduism and Sikhism, or with what we shall infer later, although Dumont will not go as far as to put caste and sect on a level as forming the warp and the woof of a single woven fabric of religion and the civil society. Brahmins, who are regarded as gods among men, possess by their birth and calling religious and social status, and lead to the three ends of law and *dharma, artha, kama*. The *sannyasis*, who are regarded as men among the gods, possess by a kind of rebirth, initiation and discipline a sacred but asocial power and virtue, and lead to the final end of self-realization or *moksha* for the individual follower. The relation of the Brahmin and the *sannyasi*, I should sum up, is bi-partitioned in medieval India between the collective and the individual, group and person, this social world and the other world, exoteric and esoteric, outer and inner, religion as a 'way of life' and religion as a 'discipline of salvation'.

The Brahmin and the king

In order to build up the structure of the elementary triad of medievalism, Brahmin, king and *sannyasi*, that is to say, the sacred or the civil society, the state and the religion of salvation, the second relation that we have to consider is that of the Brahmin and the king (Dumont 1962). The long and the short of it is that it depends upon Dumont's by now well known use of the concept of the hierarchical division and separation between 'status' and 'power'; it is a matter of both separation and hierarchy in Hinduism. 'If, in the theory of caste, the first rank belongs to the Brahman, we can say that in the actuality the hierarchy is bicephalous, except that the second head, the king, is not recognized when confronted with the first' (Dumont 1959: 85).

D.F. Pocock, his co-editor and collaborator of the first years of the journal, *Contributions to Indian sociology*, puts it plainly that the institutions of 'caste and kinship are primary in this society', the role of the political system is secondary, and 'the caste system seems to have acted against the emergence of a concept of the state, [although] it could not prevent the emergence of states'. This is still so in the Indian village, 'where the dominant caste can be seen to play the role of king'. It was so also in the Indian city at the political level. 'It is the centre of the king whose prime duty, one need scarcely stress, was the maintenance of the caste order.'

But of course this is only the case so long as Brahman and King are seen in their ideal harmony in opposition to other castes. The King (political power) is also profane and looking at the situation not now in terms of the ideal but of behaviour and history we can understand that although the Brahman's spiritual ascendancy is never challenged his *actual* power and his capacity therefore to insist upon the ideal is in inverse proportion to that of the King (Pocock 1960: 65–9).

Dumont himself concentrates rather on the conceptual framework of the relation, since

the solidarity of the first two categories, priests and princes, vis-a-vis the rest, and at the same time their distinction and their relative hierarchy are abundantly documented from the Brahmanas onwards.

For, while both the Brahman and the Kshatriya can offer the sacrifice, only the Brahman can operate it.

Similarly, we have seen that the pair Brahman-Kshatriya opposes itself, not so much to the sole Vaishya, but rather to all the rest [twice-born or not].

As the Rig Veda has already said: "He [*a*] lives prosperous in his mansion, [*b*] to him the earth bestows all its gifts, [*c*] to him the people obeys by itself, the king in whose place the Brahman goes first" (4: 50: 8, transl. Dumezil).

Concretely, the relation between the functions of priest and king has a double aspect. While spiritually, absolutely, the priest is superior, as we have just seen, he is at the same time, from a temporal or material point of view, subject and dependent. And conversely the king, if spiritually subordinate, is materially the master.

It is the combination of both aspects which actually constitutes the situation, a relation of mutual but asymmetrical dependence. . . . The situation results from the distinction between spiritual and temporal being carried out in an absolute fashion.

We can say this with some assurance, as we can observe, in the Indian villages of today, a similar relation between the Brahmans on the one hand, and the dominant caste on the other. The caste which we call dominant because it enjoys the main rights in the soil reproduces the royal function at the village level.

. . . In the matter of principle, the Brahmans as such have never claimed political power. Even in our days, they are content in essentials with guaranteeing spiritual merits to acts which are materially profitable to themselves and of which the *gift* is the prototype. To give to Brahmans is basically to exchange material goods against a spiritual good, merits (1962: 49ff, 55).

But we should carefully follow through the Hindu theoretical development, which in Dumont's hands will lead to a stratification and/or periodization in history. In the first place, Brahmin and king are separated though close, superior and inferior, encompassing and encompassed, in relation to the sacrifice.

Rather than of the first two classes, the Brahmanas (the texts called *brahmana*) treat of their principles, resp. *brahman* and *ksatra* (both neuter). They go together, they are often designated as "the two forces", and they are to be united. Similarly in Manu (9: 322) Kshatriyas and Brahmans cannot prosper separately but only in close association. But, as soon as this necessary union has been stated, the hierarchical distinction between "the two forces" manifests itself (*Pancavimsa Br.* 12: 2: 9); the *brahman* does not fall under the jurisdiction of the *ksatra*, the *brahman* being the source, or rather the womb, from which the *ksatra* springs, is superior; the *brahman* could exist without the *ksatra*, [but] not conversely.

Regarding sacrifice, the *Aitareya Brahmana* draws the logical consequence: the king must be, through appropriate rites, identified with a Brahman during the performance of the sacrifice, and be made to leave this identification at the end of the ceremony (1962: 49f).

In the second place, there emerges the implicit distinction between the role of the Brahmin as a priest (*purohit*) as against the role of the *sannyasi* as a guru, although Dumont does not make this distinction explicit.

It is not enough that the king should employ Brahmans for the public ritual, he must also have a permanent, personal relationship with one particular Brahman, his *purohita*, literally "the one placed in front".

Dumont translates *purohit* as 'chaplain', a kind of *major ego*, and then proceeds:

The gods do not eat the offerings of a king devoid of a *purohita* (*Ait. Br.* 8: 24), so that the *purohita* presides, as *hotr* or *brahman* priest, i.e. as sacrificator or controller, to royal sacrifices. Moreover, the king depends on him for all the actions of his life, for these would not succeed without him. The *purohita* is to the king as thought is to will, as Mitra is to Varuna (*Sat. Br.* 4: 1, 4).

Thus the relation between the spiritual principle (*brahman*) and the principle of *imperium* (*ksatra*) is fully seen in this institution which embodies it (1962: 51). Temporal authority is guaranteed, so to say, through the personal relationship in which it gives preeminence over itself to spiritual authority incarnated in the status of the *purohit*, who is a Brahmin, as against the other institution of the spiritual master or guru, who was a *sannyasi*.

Thirdly, the king in himself can be reduced to simply a *kotwal*, a law and order officer with a 'constabulary' function, 'the exercise of force for the pursuit of interest and the maintenance of order'; as if 'the king is just someone who is put in charge of the maintenance of public order, in exchange for which service his subjects leave to him a part of the crops they harvest'; 'the dharma texts vie with each other in repeating the formula which balances on the one hand public order, or the "protection" afforded by the king, and on the other the prestations which the king receives, and which consist first of all in a share, mostly of one sixth, of the harvested crops'. 'In politics, the agent is the Prince.' The concept of legitimate force (the army as well as justice and the police), which corresponds to the *objective* function of the prince, is combined with the interest to acquire, which corresponds to his *subjective* ends. Here *artha* denotes 'the principle of rational action directed to egoistic ends'. At the same time, 'the political aspect is reintegrated into *dharma*, as being merely its instrument' (1962: 59, 63, 65, 68).

Fourthly, Dumont brings to bear a wider comparative perspective, which I think shows that this 'contractual' theory of kingship which is modern in Europe is essentially medieval in India. It is far from and opposed to the theory of divine kingship, where the religious and the political functions are inseparably combined in the priest-king, an almost universal stage of development in human history. Perhaps something of both currents is found in India, Dumont concludes, at two different levels of society or in two different periods of history.

A priori, therefore, there seems to be a simple alternative:

either [*a*] the king exerts the religious functions which are generally his, and then he is the head of the hierarchy for this very reason, and exerts at the same time political power, or, this is the Indian case, [*b*] the king depends on the priests for the religious functions, he cannot himself operate the sacrifice on behalf of the kingdom, he cannot be his own sacrificer, instead he "puts in front" of himself a priest, the *purohita*, and then he loses the hierarchical preeminence [of status] in favour of the priests, retaining for himself power only.

Through this dissociation, the function of the king in India has been *secularized*. . . . As opposed to the realm of values and norms it is the realm of force. As opposed to the *dharma* or [all-embracing] universal order of the Brahman, it is the realm of interest or advantage, *artha*. . . . In India the king has lost his religious prerogatives . . . through a process which would have taken place in the Vedic period . . . [or] from the time of the Brahmanas until our days (1962: 54f).

Fifthly, Dumont looks at Indian myths or theories of the origin of kingship, and he finds both currents to be present. In the one, 'kingship is in some manner a divine institution'. In the Vedic and other ancient classical texts, 'the king is identified with the one or other god by reason of his nature and some of his functions'. In the Mahabharata, 'it is the supreme god who gives a king to mankind, at the request either of men or of the gods, in order to put an end to a state of anarchy and degeneracy.' Even in the Arthashastra, 'the king enjoys the earth as a wife'; thus rain and order, disorder and drought go together (*Sat. Br.* 11: 1, 6, 24), etc. (1962: 58ff).

In the other current, kingship is 'based on, or it has its origin in, a "contract" between the future subjects and the future king'. Perhaps its clearest theoretical exposition is found in the Buddhist canonical literature in Pali, where correspondingly 'group religion is banished from the tale and ultimate values only appear in individual morality'. Dumont concludes succinctly that the two currents might be found together in society in a stratified form.

. . . While the *ksatra*, or the king, has been dispossessed of religious functions proper, or of the "official" religious functions, there are at the same time, at the core of the idea of kingship, elementary notions of a magico-religious nature which have not been "usurped" by the Brahman. Below the orthodox brahmanical level, another emerges on which, certainly in contact with popular mentality, the king has kept the magico-religious character which is universally inherent in his person and function (1962: 59ff).

Alternatively, one may suppose that the Indian situation corresponds to two different periods of history.

From our sources, this seems to have developed in two stages: the first stage, I have tried to show, was attained very early, within orthodox brahmanism, in the relation between *brahman* and *ksatra*; the second stage appears to have been the work of a current which was non-brahmanical in its orientation, the thought of individualists, i.e. of renouncers.

The first event, which really sets the stage for Indian history, is the secularization of kingship laid down in the *brahman-ksatra* relationship. It invites us to revise some current notions about the relation between hierarchy [status] and power. The second event, or stage, is more complex. It has appeared to us under two forms: on the one hand in the idea of contractual kingship, which appears to emanate from renouncers [Buddhist or *sannyasi*], on the other in the theory of *artha*, not unconnected with the renouncers' individualism and their negation of brahmanical values, but constituting a politico-economic domain.

This domain is, in the dominant tradition, *relatively* autonomous with regard to absolute values. In so far as it is autonomous, there is at this stage a rough parallel with the modern western development, and this leads to a generalizing hypothesis, namely that such a domain as we know it necessarily emerges in opposition to and separation from the all-embracing domain of religion and ultimate values and that the basis of such a development is the recognition of the individual.

I should have said instead rather a recognition of 'the individual *and* the nation'.

To stress the point, let me anticipate on another study and say that in the West, the political sphere, having become absolutely autonomous in relation to religion, has built itself up as an absolute: comparatively, the modern "nation" embodies its own absolute values. This is what did not happen in India (1962: 63, 75f).

The sannyasi *and the king*

Although he had promised to speak as far as possible a 'language of relations' (1960: 34), Dumont has nothing to say about the relation of the *sannyasi* and the king, except for the single remark that the *sannyasi*'s unique position gave him 'a sort of monopoly for putting everything in question' (1960: 47). This is the one relation that is not reducible to the logic of hierarchy, I suppose, the encompassing and the encompassed, a gradation of statuses or more specially the 'encompassing of the contrary', which so fascinated Dumont in the world of the Brahmin that the total underlying structure of medieval India escaped him.

Left to our own device, therefore, I may say that this apparently invisible relation is the moment of conjunction of religion as the 'discipline of salvation' and the power of order or the order of power. It is the very point or fulcrum of medievalism from which, for example, Sikhism was the first and Gandhism the latest, not to say the last, attempt to move oneself and India from medievalism to modernity, i.e. from dualism to non-dualism as I should define it, and without denying the permanent Indian inheritance. The sacred social status of the Brahmin, supposedly all-embracing, and the political power of the prince, executive and politico-administrative but not legislative, were both of the collectivity and also exoteric in nature. The medieval role of the *sannyasi*, on the other hand, was manifestly to bring both status and power together under spiritual mastery, but only at the expense of himself moving or being moved from the collective to the individual, the exoteric to the esoteric, asocial and apolitical. The renouncer could heal others as well as himself, one has to admit, but not society or the polity, not to say the economy, in medieval India. For that final revolutionary step to the Indian modernity required a Platonic confrontation and marriage of the philosopher and the king, a consummation that was to be achieved, not by setting up an Utopia, but through the example of martyrdom versus kingdom, as we shall see later.

I shall try and show in the following two chapters the project of Sikhism set to produce a society for salvation and self-realization that should so reconcile in the non-dualist Indian modernity the two intersecting axes (*a*) of status versus power, on the one hand, and (*b*) of the collective versus the individual, exoteric versus esoteric, on the other hand. My proposed investigation of historicism, it may be said in advance, is not a claim of success, either for Sikhism or for Gandhism, but a claim of the project to provide a necessary and sufficient example, worked out to hold truth in a dialectical relation with reality, transcendental and immanent.

Meanwhile, I do not wish to place any kind of blame on the elementary structure of medievalism in India, symbolized by the trinity of the Brahmin, the king and the *sannyasi*, which we have resolved following Dumont into the two underlying bi-partitions between the collective and the individual, and between status and power. Indeed, unlike Dumont, one might argue from the evidence that, in the view of a comparative sociology of religion, the same sorts of dualisms, far from being specific or unique distinctive features of Hindu India, a tradition showing to him an 'extremely remarkable permanence' (1962: 48), were an almost universal stage in the history of humanity. One can immediately cite in support of this other view the well-known paper by Bellah (1964) on religious evolution to the same effect below.

Bellah's scheme of religious evolution, ideally intended to be part of a general theory of social evolution, is in five stages, (*a*) primitive, (*b*) archaic, (*c*) historic, (*d*) early modern, (*e*) modern. The middle type, historic religion, corresponds to what I have called medievalism; its study has fallen chiefly 'under the discipline of history rather than that of archaeology or ethnography'. 'The dualism that was so crucial to all the historic religions', Bellah holds, is the concomitant of their transcendentalism, so that the religious goal of salvation, liberation or enlightenment lies in making a separation and difference between man in this world and God, life after death, the true self and the other world. 'The historic religions discovered the self'; and incidentally opened the possibility of 'remaking the world to conform to value' (1964: 361, 366, 371f, 374).

Bellah therefore begins with the emergence in the first millennium BC of the 'phenomenon of religious rejection of the world' all across the Old World, and the exaltation of the other realm of reality as alone true and the source of value. This 'dualist' theme emerges simultaneously in Greece, Israel, India and China; it is characteristic of a long and important period of history; and contrasted with its virtual absence before that in primitive religion and after that in the modern world.

For over 2000 years great pulses of world rejection spread over the civilized world. . . . Even in Japan, usually so innocently world accepting, Shotoku Taishi [d. 621 AD] declared that the world is a lie and only the Buddha is true, and in the Kamakura period [1185–1392] the conviction that the world is hell led to orgies of religious suicide by seekers after Amida's paradise. And it is hardly necessary to quote Revelations or Augustine for comparable Christian sentiments.

The net result of this development was that:

The profound dualism with respect to the conception of reality is also expressed in the social realm. The single religio-political hierarchy of archaic society tends to split into two at least partially independent hierarchies, one political and one religious. . . . The differentiation between religious and political that exists most clearly at the level of leadership tends also to be pushed down into the masses so that the roles of believer and subject [citizen] become distinct. Even where, as in the case of Islam, this distinction was not supported by religious norms, it was soon recognized as an actuality.

The differentiation of a religious elite brought a new level of tension and a new possibility of conflict and change onto the social scene. Whether the confrontation was between Israelite prophet and king, Islamic ulama and sultan, Christian pope and emperor or even between Confucian scholar-official and his ruler, it implied that political acts could be judged in terms of standards that the political authorities could not finally control (1964: 359, 367f).

From my point of view, then, we find (*a*) that the historic religion as a type is, on the one hand, the beginning of the end of the institution of the divine kingship, the figure of the priest-king as the chief link between the people and the gods, and between religion and politics, a general tendency of earlier societies. On the other hand, it appears (*b*) that the 'individual and his society' are no longer seen as merged together within a 'natural divine cosmos', but as mutually contrasted and opposed, as I have shown them in the elementary structure of medievalism. Bellah comes close enough to sense this second dimension, if it is not unique to India, but his pure sociologism of interest leads him to emphasize a different conclusion, rather like that of Dumont.

The emergence of the historic religions is part of a general shift from the two-class system of the archaic period to the four-class system characteristic of all the great historic civilizations up to modern times: [*a*] a political-military elite, [*b*] a cultural-religious elite, [*c*] a rural lower-status group (the peasantry) and [*d*] an urban lower-status group (merchants and artisans).

He correlates this social evolution with other factors such as the growth of literacy, the market, bureaucracy, law and urbanization (1964: 365, 367).

The student of pre-Islamic Iran under the Sassanids, for example, will immediately recognize something like the four-class system of (*a*) the 'theologians' or priests (*atharvan*), (*b*) the warriors (*rathaeshtar*) and officers of government (*dabiran*), and the common people, including (*c*) agriculturists or the peasantry (*vastriya*) and (*d*) traders and artisans (*huiti*). It is not known whether such a social organization was among the ancient institutions of civilization which the early Aryan settlers brought with them and matured in Iran and India.

The Muslim Culture of Medieval India

The sociology of Islam, in a strictly professional sense, is not so well developed as a discipline, in spite of all the important work done since Levy's two volumes first used that title in English in 1931 and 1933, so that it is to the discipline of history that we must turn now. In his introduction to *Some aspects of religion and politics in India during the thirteenth century*, Mohammad Habib of the Aligarh Muslim University will be a helpful guide to the synchrony as well as diachrony of the Muslim trinity of *shari'at, tariqat* and *hukumat* in the middle ages. So far as the Orientalists are concerned, whom we must set aside for the moment, the corresponding languages, or even races, of the realm of Islam, are the Arabic, the Persian and the Turkic, or more accurately,

shari'at : *tariqat* : *hukumat*
::
Arab : Iranian : Turk.

Shari'at *and* hukumat

Habib of Aligarh says that the military invasions of Sultan Mahmud 'only brought disgrace to the creed he professed'; and that neither in Iran nor in India has modern research been able to find any authentic contemporary literature to show how Islam

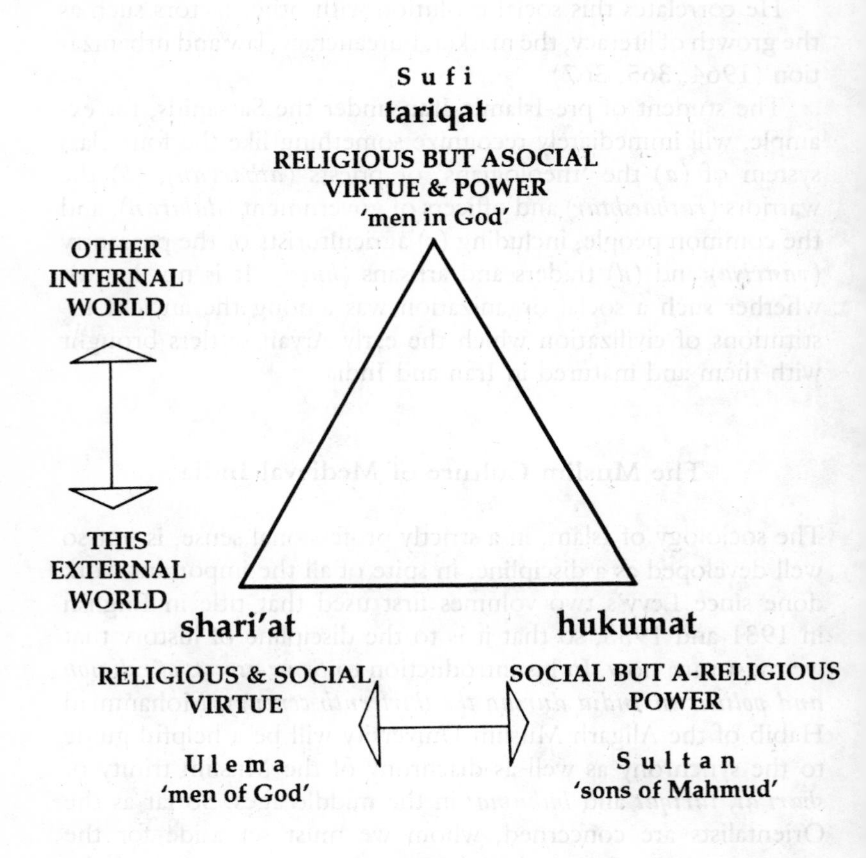

FIG. 2 The Muslim Culture of India

spread in either country (1961: xiii). Nevertheless, we also have it on the very good authority of W. Barthold, for example, that the concept of the state was brought to its 'extreme' expression under the Ghaznavids, whose possessions straddled the Indo-Iranian frontier, and specially under Mahmud, whose advent and reign as the first sultan of 'Ajam, the non-Arab lands, is reckoned to begin a new epoch in the history of Islam (999–1030). Mahmud of Ghazni's raids into India, beginning in AD 1001, and specially his idol-breaking exploits at Mathura and Somnath, which might be partly apocryphal, did not accomplish much, but he did assure the permanent accession of the Punjab, or at any rate of half of it, to the realm of Islam.

Mahmud of Ghazni, followed by the Turkic dynasties of the Ghurids and the sultans of Delhi, followed in turn by the Mughals, all tried to combine in the person of the ruler, sultan, amir, king or emperor, but without apparent success, the traditions of Baghdad and Bukhara, west Asia and central Asia, roughly corresponding to the dualism of status versus power in Islam.

The caliph of Baghdad had granted the patent of investiture (*manshur*) to Sultan Mahmud in AD 999, for 'no king of the east or west could hold the title of *Sultan* unless there was a covenant between him and the *Khalifah*' (Khalil bin Shahin al-Zahiri). In Islamic law, therefore, Mahmud was the lieutenant of the caliph, who was the lieutenant of the Prophet, who was the lieutenant or the deputy of God, although the Quran uses the term *khalifah* of Adam and not of Muhammad. But, like his own former overlords, the Samanids of the two central Asian capitals of Bukhara and Samarkand, Mahmud himself 'strutted as an absolute autocrat' (Nizami 1961: 26, 122). He was the independent politico-administrative master of 'Ajam, its sultanate or empire, the 'shadow of God on earth'; and he claimed, the historian says, to exercise supreme legislative, judicial and executive authority. On the other hand, the Abbasid caliphate retained its being as the formal symbol of the political as well as the religious supremacy of Islam, until it was extinguished by Hulagu the Mongol, who sacked Baghdad in 1258.

During the time of the Prophet, and in the language of the Quran, the word *sultan* usually meant 'authority' in the spiritual rather than the temporal sense of either power or argument.

There are half-a-dozen places in the text of the Quran where the term *sultan* occurs with the negative spiritual meaning of the power that Satan or Iblis exercises or can exercise over men (14: 22, 15: 42, 16: 99, 100, 17: 65, 34: 21). A seventh similar usage relates to him who will be given his record in his left hand on the last day; he will say, 'My power has gone from me' (69: 29). In the early centuries after the Prophet the term came to mean the governmental power, and so eventually became a personal title of the ruler (Nizami 1961: 95n).

Mohammad Habib sums up the Muslim development for us:

[*a*] The Prophet left the organisation of political and administrative affairs to the secular good sense (*ijma'*) of his community. [*b*] Amir Mu'awiya changed the Caliphate into a *mulk* or monarchy, though he continued the name. [*c*] With the rise of Ghaznavid power the first Sultanate in Islamic history was established.

But the sultanate had no sanction in the *shari'at*; indeed, it was not a legal institution of Islam. Its laws were a result of the legislative activity of the executive rulers and the governing class, whose institutions were borrowed from the Samanids by the Ghaznavid rulers at Lahore and passed on to the Ghurids and, after the defeat of Prithvi Raj in 1192, the sultans of Delhi. It is the political system of the Samanids that is described by Nizam ul-Mulk's *Siyasat-namah*, treatise on government (*c.* 1092).

The absolute monarchy here described lasted as long as hereditary rulers were tolerated among the Musalmans—in India right till the death of Aurangzeb; and one of the elements that contributed to its maintenance—fear of anarchy—still expresses itself in present-day Muslim preference of autocracy to democratic regimes (Habib 1961: viii).

Even an hereditary absolute monarch, however, could not rule without placing his reliance on the officer corps, Muslim nobility and bureaucracy, let alone the Hindu merchants and princes. This ruling class was formed of men originating from the Turkic-speaking parts of eastern Iran, Khurasan, Tukharistan (Balkh) and Transoxiana. Habib defines it roughly as comprised of people who had been living north of a line joining the gulf of Alexandretta to southern Badakhshan. After the accession at Baghdad of al-Mu'tasim, caliph from AD 833 to 842, the Turks who had been brought in as the Caliph's bodyguard proved stronger than any other group. Meanwhile the Samanid rulers [*c.* 874–999 or 819–1005], who were Persians in origin but governed a predominantly Turkish population, carefully organized the Turkish slaves for the service of the state. As a result of these two movements the higher as well as second rate offices came into the hands of various Turkish groups in all Muslim countries (1961: ix).

Turkish rule was overthrown in Iran after six centuries by Ismail Safavi, who also made Shiah Islam the state religion in 1501, but in many other Muslim lands it lasted until the rise of modern European imperialism. In India, Habib writes that it was exceptional for a non-Turk, Hindu or Muslim, to occupy high office.

The whole responsibility of the policy of the government rested upon these Turks or Turanis. During the Moghul Empire they formed a closed group of ***mansabdars*** called *Khanazadas* [70 per cent].

On the other hand, the Hindus who held appointments numbered but 15 per cent or less; and of the remainder it is said,

These [Indian-born] Musalmans and the Hindu converts who joined them [together 15 per cent or more] had no say in the ***policy*** of the government of medieval India.

Altogether, then, the administrative unification of India was a Turkish achievement, and this chiefly because 'the average Turkish governor realised that without a centralised authority, he would be crushed between the Hindu rulers and the Mongols' (1961: xii, xix).

Thus the theocracy of Islam, properly so called, lasted for scarcely thirty years after the death of the Prophet in AD 632, i.e. the short classical period of the four 'rightly-guided' or patriarchal caliphs. The first seat of the caliphate was al-Medina, 'the city', and then, already under Ali, the camp-city of Kufa. Mu'awiya, first caliph of the Umayyad dynasty which followed in 661, moved the capital to Damascus; under the Abbasids, who succeeded in turn (750), it moved to Baghdad until 1258. The modern Ottomans later claimed, after Mongol rule had subsided, that the last Abbasid descendant had ceded the title to Selim I upon his conquest of Cairo (1517). Then Istanbul became the capital of political Islam as well as responsible for defending religious orthodoxy until the caliphate was formally abolished as an institution in 1924.

In the beginning, when its territory was rapidly extended from the Indus to the Atlantic, the worst critics of the institution of the caliphate were the Kharijis, sectarians of theocracy, who saw no need for it as a substitute for the office of the Prophet charged with the dual functions of defending the faith and administering the worldly affairs of the universal Muslim empire. They held the view, which is in itself perfectly orthodox, that after his death the mantle of the Prophet, if not as messenger of God, then as patriarch, priest and prince, had fallen upon the community of Islam as a whole (*ummat*), and not upon any separate priesthood, hierarchy or authority, but they wanted to push the matter to its logical—or illogical—conclusion, and with violence.

> Ali in the end met his death at the hands of a group of fanatics, the "Kharijis", or "Seceders", who cared neither for him nor for Mu'awiya but regarded both as usurpers, on the grounds that neither was ever raised to the chieftainship of Islam by that free choice [of the community] which Arab custom demanded. These democratic Arab tribesmen detested both the sanguinary family rivalries of the aristocrats of Mecca and what they considered the hierarchical pretensions of those in power at Medina. They, in fact, disputed any need at all for any imam, [caliph] or head of the State, as long as the divine law was carried out, and they felt that the victory of either of the two rivals [Ali and Mu'awiya] would mean the triumph of worldliness over the religion of Islam (Levy 1957: 278f).

In the end, the caliphate of the Ottoman empire, where the law was based in theory on the interpretation of the *shari'at* according to the Hanafi school as in India, was dismantled by the secular modernists precisely because, as the preface to the decree of 1924 says, 'the Caliphate and the Sultanate were two distinct authorities united in the same person'—but united unsuccessfully and/or fictitiously. There resulted in fact two separate jurisdictions, one regarded as belonging to the representative of the sultanate (the grand vizier), and the other to the representative of the caliphate (the Shaikh al-Islam), but who regarded himself as independent. It came to be seen that this position of affairs developed a 'State within the State', and so the National

Assembly determined to subject all law, civil, personal and criminal, to its own sole control. This automatically also meant the abolition of privileges held until then by the independent ulema, doctors or legists of the *shari'at*.

The term used for the making of laws in the decree is *tashri'* (i.e. the formulation of *shar'*), which hitherto had not been used except in speaking of the law-making of the Prophet (Levy 1957: 269).

A formal statement of the problem of the caliphate, then, will help to clarify the matter of religion and politics as it was understood in medieval India. The caliph had always the status but scarcely or never the power to defend the faith, which was his legitimate responsibility, and conversely administration of the affairs of a Muslim land was in the absolute power of an amir or a sultan who had no status in theory beyond that of an executive officer or a magistrate. By defending the faith was meant the daily tasks of 'commanding the right and forbidding the wrong' at home as much as guarding the frontiers and outlying regions of the territory of Islam. On the one hand, we have (*a*) the ulema, who hold religious and social status, men of God, who are chiefly the learned of the sacred law in orthodox Islam; on the other hand, there are (*b*) the 'sons of Mahmud', as the historian Ziyauddin Barani called them, who hold and represent social but a-religious power, i.e. the state (*hukumat*). Between the two there developed a dual legal system, religious *versus* politico-administrative, in relation to its sources, institutions and application, although the executive arm always prevailed in practice or tried to do so. The sacred law was reduced to matters of individual piety, ritual (prayer, fasting, pilgrimage, alms or poor-rate) and personal law (e.g. marriage, divorce and inheritance). Conversely, the law and administration of the collective life of Muslims depended upon two non-canonical sources, imperial will and political expediency (*siyasat*) and local custom (*'urf* and *'ada*), which could never be recognized as valid by any of the orthodox schools of law.

So far as legislation for the community of Islam is concerned, [even] the Caliph had in theory no powers at all, seeing that God is the only lawgiver and that he had declared his will in the Koran as revealed by the Prophet Muhammad. Any seeming obscurity or inadequacy in the

Koran was made good by the *hadith*, which was the record of the Prophet's [example,] sayings and doings and was a sufficient guide to the will of Allah. Failing them, and where there remained still some uncertainty or some need for a new application of the revealed law, a number of authorities, namely, the imams, the founders of the four "schools" of law, interpreted or expanded the laws as they stood.

The Caliph was not endowed with any special privilege in this respect. He was in theory merely the representative and agent of the law, the person by whose efforts it was carried into effect [the executive]. He had himself to observe it as well as to secure its being observed by others.

On the other hand, once the religion of Islam came to be interpreted almost exclusively by the *'ulama* and the legists, their influence over the community overruled that of the Caliph himself, and when the Caliphate disappeared [1258] the sovereignty often lay in their hands (Levy 1957: 294f).

The science of *shari'at* defined as the sacred law, path and road of Islam properly includes both the science of *fiqh* (jurisprudence), 'knowledge of the practical rules of religion', and the science of *kalam* as dogmatics and scholastic theology or what we should call philosophy of religion. Everywhere in Islam, it is to be noted,

there has never been any dispute about such [religious] questions as the unity of God and the eternal validity of the Koran. The Shi'ites may part company from their "orthodox" brethren on such serious questions of political theory as the succession to the Prophet, but all four Sunni "schools" are at one over these. Where there are differences between them, the points at issue are seldom of practical importance. Much indeed of the whole body of *fiqh* is detached from reality, for though all books of law purport to regulate not only the acts of the individual Muslim but also the governance and transactions of the state, yet ruling princes and conquerors have generally applied such methods as [political] expediency dictated rather than those which [religious] authority demanded.

Consequently only the ritual and personal laws and those dealing with the administration of *awqaf* [sing. *waqf*], or pious foundations, have in fact had consistent practical application, and the rest of the *shar'* has been regarded by those religiously engaged in studying it as an ideal to be brought into actual use only at the coming of the Mahdi, the last precursor of the Resurrection (Levy 1957: 185f).

The great imam Ghazali (d.1111), however, felt compelled

to admit that in the middle ages the acts of the secular state and its administration were valid in view of 'circumstances of the time'. There was also some attempt later to believe that, with the fall of the universal caliphate at Baghdad, its functions were to be considered as redistributed in severalty, so that every independent Muslim ruler 'should discharge the duties of a caliph inside his realm'; and to combine this with the contractual theory of kingship by a kind of local election process.

The sultans of Delhi adhered to the legal conception of the position of the sultan which was common throughout the Muslim world. They also adhered to the form of an election by the *elite.* After the nobles had formally elected a monarch, they swore allegiance and later, the oath was taken by the people in the mosques of the state. The election of a sultan was in fact often purely nominal, because the candidate had already decided the issue by conquest or by the possession of superior force. The sultan was virtually bound by the election [*a*] to control the state, [*b*] defend Islam and its territories, [*c*] protect his subjects and settle disputes between them, [*d*] collect taxes and rightly administer the public treasury, and [*e*] enforce the criminal code.

It follows that an elected sultan could also be deposed for breach of those conditions, and a number of sultans of Delhi were in fact so removed for incompetence. . . . As in other Islamic states, the mention of the ruler's name [along with that of the caliph] in the *khutba* [Friday congregational prayer] and the minting of coins in his name [along with that of the caliph] were considered the most important attributes of sovereignty (*khutba-wa-sikka* in Qureshi 1970a: 30).

The Mughal emperors claimed [*a*] to be fully independent monarchs, and [*b*] to be caliphs within their dominions. . . . The monarch was the chief executive of his realm, and the commander of its forces. His power was limited by the *Shari'a* (Qureshi 1970b: 52).

The medieval Indian position in theory and practice, which has been studied by Nizami (1961), for example, shows that the dualism of status and power underlies much else besides the relation of the caliph and the sultan. I refer specially to the relation of the latter to the ulema, and to a kind of induced split within this class of doctors of the law, who are the educated and learned of Islam.

In the first place, we may accept Nizami's conclusion that 'all Muslim governments from the time of the Umayyads have been secular organizations'; and that the 'administration was exclusively in the hands of the secular authorities'. All through the middle ages, the view expressed by Muslim writers on political ethics was the same as the one quoted in Nizam ul-Mulk's *Siyasat-namah* saying that government can be carried on successfully by just infidels but not by unjust believers.

Militarily, this would imply, for example, that it was population pressure at home rather than religious fanaticism that had brought the Turkic-speaking groups into Hindustan; and similarly that the non-Muslim Mongol pressure had driven Muslim families pell-mell from central Asia to supply man-power for the infant Turkish empire of India.

> The Turkish invasions were not inspired by any religious zeal or proselytizing fervour. Shihabuddin's first conflict on the Indian soil took place not with a Hindu raja but with a Muslim co-religionist, and he faced him [the Ismaili heretic of Multan, 1175] with the same determination and in the same spirit in which he carried his arms into the *Aryavarta*. The Ghurid successes were not followed by any retaliatory action inspired by religious zeal or fanaticism (1961: 87ff, 111).

Secondly, again following Nizami, one can divide the social structure into the two-headed elite versus the non-elite public or the common people, whose opinion had no recognized or regular institutions of expression. On the one hand,

> No Sultan could ignore the nobility and the *'ulama*; if the one controlled the administrative machinery, the other controlled the public opinion.

On the other hand, one can look at the same relation of the sultan and society from the reverse side:

> [*a*] The *'ulama* wanted him to champion the cause of religion; [*b*] the governing clique wanted him to act as the guardian of their political interests and privileges; [*c*] the common man expected peace, security and justice from him. . . . It may be safely stated that the people could accept and tolerate any Sultan provided he guaranteed peaceful conditions and administered even-handed justice (1961: 111f).

Thirdly, the opposition of religion and the world, *din* and *dunya*, status and power, was everywhere to be seen and to my inference unreconciled.

If the *fatawa* collections [of decisions on points of law, sing. *fatwa*] of the middle ages are scrutinized as a whole, it would appear that while innumerable problems relating to civil, personal and religious matters have been discussed in detail, there is hardly any reference to political or administrative problems. . . . In administrative matters the Muslim governments were guided, not by the *Shari'at* laws, but by what [Ziyauddin] Barani [1285–*c.* 1357] calls the *zawabit*, i.e. the secular regulations framed by the rulers in the light of the exigencies of the time [the Mughal *qanun-i shahi*].

Even the ethico-political treatises of the period [e.g. *Fatawa-i jahan-dari*], though anxious to invest the king with divine dignity, did not hesitate in declaring the incompatibility of *din-dari* with *dunya-dari* of which [latter] the Sultanate was the highest perfection.

Since no sanction for the Sultan or the Sultanate [versus the caliphate] was available in the [normative] laws of *Shari'at*, the contemporary writers tried to justify it on grounds of [empirical] *necessity* and said that "if there were no king, men will devour each other".

Participation of the *'ulama* [sing. *'alim* or mullah] in political matters was deemed harmful and improper. Ibn Khaldun [1332–1406] considered them utterly incompetent to tackle political problems due to their ignorance of the requirements of time (Nizami 1961: 40, 90, 110, 151).

Fourthly, it is true that one may find in medieval India some apologetics for what has been aptly called 'a partnership between pious professors and pious policemen' (Hardy 1958: 465), i.e. a partnership between the ulema and the sultan or the nobility, so as to better serve the higher common interest, but it was never quite certain as to what this latter was. The orthodox indeed feared that it might prove to be a reversion to the pre-Islamic and ancient traditions of Iran and Rome, the practices of Khusrau and Caesar, when religion was regarded as the sister of kingship, which is the way Abul Fazl would have it under Akbar, instead of the other way about. The common people, it seems to me, always preferred their ulema to remain independent albeit largely ineffective in relation to affairs of state; and correspondingly looked to the sultan or amir for assuring peace, prosperity and justice rather than for seeking either the way of spirituality or glory in war. In many traditions of the Prophet, it was said, the ulema are referred to as his own heirs and sometimes compared to the prophets of the Israelites, a status higher than that of any sultan, amir or nobleman (I suppose).

According to a contemporary historian of the middle ages, 'the *'ulama* are much superior in dignity and status to others. After them rank the kings.'

The Prophet has said: "The best kings and the best nobles are those who visit the doors of the *'ulama* and the worst *'ulama* are those who visit the doors of the kings and the nobles" (*Tarikh-i Fakhruddin Mubarak Shah*, AD 1206, transl. Nizami 1961: 150).

Fifthly and finally, (*a*) the two opposed faces or orientations of status and power are to be seen manifest, back to back, in the internal division occurring within the medieval ulema as a category or class; while (*b*) the other axis of medieval Indian dualism, the individual versus the collective or esoteric versus exoteric, is made manifest within the person of even the best pious sultan, e.g. during the reigns of Iltutmish (1211–36) and Balban (1266–87).

'Tradition classified *'ulama* into two categories—the *'ulama-i akharat* and the *'ulama-i dunya*', the ones mindful of the last day or the day of judgment and the ones mindful of life in this present world. 'The basis of this division was the difference of their attitude towards worldly affairs.' The first category led an abstemious life of piety, devotion to religious learning and teaching in the mosque or at home, eschewed materialistic pursuits and political affairs and often lived in penury, 'unspoilt by wealth and power', trying hard to preach Islam, disseminate knowledge and strive for the moral uplift of society. The other category was totally 'mundane' in its outlook, aspired for wealth and prestige, mixed freely with kings, nobles and the bureaucracy, both sought and accepted from them offices and benefices (*sadr-i jahan* or *shaikh al-Islam, qazi, mufti, muhtasib, imam* and *khatib* or a teacher in a state school or *madrasa*), and in return 'gave them moral support in their actions, good or bad' (Nizami 1961: 152ff).

Muslim public opinion, we are told, also called them *'ulama-i su*, a thing of evil, and treated them with contemptuous indifference, or else like the later Ahmad of Sirhind (1564–1624), 'renewer of the second millennium of Islam', shaikh of the Naqshbandi order, held them responsible for all the vices and misfortunes of civil society.

Shaikh Nizamuddin Auliya of Delhi (1238–1325) is said to have praised in his life only two medieval kings for their religion, Iltutmish and Balban, and contemporary orthodox opinion was surely at one with him, but both rulers evidently successfully separated their personal religion from their political office as the individual from the collective.

Iltutmish had somehow been in his youth at both Baghdad and Bukhara, where he had 'imbibed an esoteric spirit of religion', but Nizami has to concede that, 'whatever his personal religious outlook, he refused to take any [official] risks by ignoring the demands of political life' (1961: 97, 116).

Balban, who was also a native of Turkistan and had been in Baghdad as a young captive, was on the best of terms and intervisited with the ulema as well as the Sufis in Delhi. A modern Western historian has said of him that 'no one understood better than Balban the conditions of kingship in India' (S. Lane-Poole). Thus, one can see the two roles of religion and politics subdivided in his person and rule:

Balban evinced great interest in religion so far as his personal life was concerned, but, as Barani very correctly remarks, he never cared for the laws of *shari'at* in dealing with those who defied his authority or who were found guilty of any political crimes (*Tarikh-i Firoz Shahi, c.* 1357, Nizami 1961: 98, 120f).

Shari'at *and* tariqat

The Islam of the ulema, we may say, is everywhere the transcendental religion of revealed tradition, law and the community, even if it includes much else besides on its syllabus. It concentrates on the serried ranks of the faithful at prayer in congregation arranged, as it were, like Ibn Khaldun's preferred order of battle (the analogy is his in writing on the subject of war and organization, see Levy 1957: 454f). This aspect of Islam is standard as well as exoteric, and prescribes the minimum practical 'code of conduct' for the individual Muslim in his or her external relations to God, man and nature. Sufi Islam or the mystic way, on the other hand, is said to be immanentist, and it always puts first spirituality and the relation of man or woman to oneself, the true self of the inner God. It begins, therefore, not with the fact of birth in a Muslim family or society (objective), but with the experience of rebirth or initiation by choice (subjective) into a mode of thought and a 'discipline of salvation' (*tariqat*) marked by stages of self-realization under the guidance of a pir or shaikh in a private retreat of worship and contemplation, or an organized convent of dervishes (*khanqah*). In either case, Sufism is normally represented as being esoteric and essentially individual in character, as when each neophyte gets his or her own spiritual formula whispered in the ear for repetition and invocation. The Sufi order, sect or convent may be organized, with a set text, method, technique and routine, but the discipline is believed lifeless without the central personal leader/follower relation. Even the best ordered sects usually scrupulously avoid setting up another society of private property, occupation, production and reproduction (yet with some 'communist' exceptions to which we shall have to refer later on).

The complementarity as well as the opposition of these two aspects of Islam, the transcendental and the immanentist, is everywhere recognized in some form but not to my mind always reconciled in theory and practice. Ghazali (1058–1111), known in himself as the 'proof of Islam', who was a Persian, suitably distinguished between the *'ulama-i zahir*, externalist scholars who proceed from 'knowledge to action', and the *'ulama-i batin*, saints and mystics who proceed from 'action to knowledge'. Ghazali best formulated the great medieval synthesis of tradition (*naql*), intellect (*'aql*) and spirituality (*kashf*), it is said, by adopting a Sufi method to realize, rather than to negate, orthodox ideals. The more usual way, however, is expressed by the Turkish saying about the three stages of self-realization proceeding in the opposite direction, i.e. from *shari'at* or the law, 'yours is yours and mine is mine', to *tariqat* or the way, 'yours is yours and mine is yours too', to *ma'rifat* or final gnosis, where 'there is neither mine nor thine'.

The relational opposition of what is *zahir*, i.e. clear and evident, manifest or apparent, literal, external, exoteric, and what is *batin*, i.e. hidden, spiritual, internal, secret and esoteric, is of very general importance in Islam. It corresponds in the Quran to the opposition in time of the first (*awwal*) and the last (*akhir*); and both pairs of terms are employed as names or attributes of divinity, 'He is the first and the last, and the manifest and the hidden, and He is all-knowing of all things' (57: 3).

In a way, it also corresponded in India to the bi-partition or classification of the internal division between two kinds of Sufi faqirs or dervishes. (*a*) The *ba-shar'* sects and orders who remain within the Islamic law and follow a code of conduct according to orthodox principles; they have regular links with the civil society through the institution of the shaikh, rather like a guru, his lay followers and the *waqf* or pious foundation. (*b*) The *be-shar'*, without the law, who are taken to be anti-social because they apparently do not rule their lives according to any external discipline, individual or collective, although they call themselves Muslims. Members and followers of the former are called *salik*, pilgrims or travellers on the pathway to heaven, whether their *silsilah* or chain of succession is Qadiri of west Asia, Suhrawardi, Chishti, Jalali of central Asia, Shattari or Naqshbandi, etc., so that now 'there is scarcely a maulawi or learned man of reputation in Islam who is not a member of some religious order'. The latter are called either *azad*, free, or *majzub*, i.e. whom God has chosen for himself and abstracted from the world, and, like the *qalandar* or wandering dervishes of the *Thousand and one nights*, they 'can scarcely be said to be Muhammadans, as they do not say the regular prayers or observe the ordinances of Islam' (Hughes 1895: 116; see also Glassé 1989: 321).

Personally I concur with the judgment expressed by Fazlur Rahman that Sufism has constituted the biggest challenge to Muslim orthodoxy down to the dawn of modern times, although it is, at any rate to begin with, 'incurably individual' and it often ran counter to the orthodox ethos of the community. In the view of the ulema, of course, 'society is not a venue for individual self-realization', but rather for conformity to the sacred law and the divine will (Hardy 1958: 509). Nizami is also persuaded that it has been 'through the mystic channels alone that dynamic and progressive elements have entered the social structure of Islam', even if the mystic himself was expected to reject the world.

For its part, the Sufi seminary, retreat or convent would often receive from a worldly source a pious endowment for its maintenance (*waqf*). If it did not, then its shaikh or pir would instruct the neophytes, as against the lay followers, to earn their livelihood outside, or permit them to go and beg for alms, or ask them to sit in the *khanqah* resigned to God's will (*tawakkul*). Sufism was not without its social causes and effects, therefore, but no one will dispute that its individualism and esotericism, viewed as its distinctive features, separated it from the polity and the economy of the world.

> . . . The Sufi movement started in the second [AH]/eighth [AD] century, partly [*a*] as a reaction against the political situation, and partly [*b*] as a complementary antithesis to the development of the systems of law and theology in Islam. . . . But Sufism, like Shi'ism, threatened to drift from the social and communal ethos of orthodoxy, both [*c*] by making the individual the centre of its attention, and [*d*] by its doctrine of esotericism (Fazlur Rahman 1970: 633).
>
> If its metaphysics attracted the higher intellects, the mystic ceremonials—*sama'*, *'urs, langar,* etc.—drew to its fold the common man who looked upon the mystic more as a blessed miracle worker than the teacher of a higher morality (Nizami 1961: 50, 62, 237).

At the same time and during the same period, it is to be remarked, the orthodoxy of the ulema was suffering from 'over-organization', as Nizami aptly terms it, marked by exaggerated respect for the past, and free thought was muzzled, whether out of fear of anarchy and schism or civil war, or from a preference for solidarity and uniformity (unity) versus purely local systems (variety). After the death in AD 855 of Ibn Hanbal, founder of the last of the four orthodox schools of law, the 'gate of independent reasoning' (*ijtihad*) was considered as closed, and by 1200 the time of independent interpreters (*mujtahid*) of even the third degree was ended by learned consensus. Since then no one has been recognized as qualified for the 'effort' required to open it again, although, for instance, Abul Fazl's candidate, the illiterate emperor Akbar, pretended to secure such recognition from some ulema at his court in 1579 (Abdul Qadir Badauni).

> The religious history of Islam after the twelfth century, particularly in those lands of the Eastern Caliphate which later came under the political dominance of the Turks and the Mongols, was largely that of the Sufi mystic movements and of the struggle of the ulama to keep those movements within the Muslim fold (Hardy 1958: 411).

According to Nizami's showing, the pedagogy and syllabus

of medieval Indian institutions, whether supported by government or private, reflected no innovation (and no thought on the science of nature, although this latter inference is contested by A. Rahman). Nevertheless, the Indian syllabus was catholic, and it certainly included *tasawwuf*, the science of mysticism, even if it probably drew the line at Ibn Arabi. The list of studies consisted of Quranic exegesis (*tafsir*), traditions of the Prophet (*hadis*), Islamic law or jurisprudence (*fiqh*), principles of Islamic law (*usul-i fiqh*), Sufi mysticism (*tasawwuf*), literature (*adab*), grammar (*nahw*), scholastic theology or philosophy of religion (*kalam*) and logic (*mantiq*) (1961: 151n).

To sum up its relation to the world and trying to combine both its synchrony and diachrony, Sufism might remain within the fold to serve as the leaven or the safety-valve of Islamic orthodoxy, and linked to the civil society through the relation of pir or shaikh to his *murid*, initiated follower or lay disciple, as well as through the system of *waqf* or the pious foundation. (The principle of distribution of territories or *wilayat* among saints and orders is not yet sufficiently well understood by us.) Alternatively, however, the development of orders and sects might take the path of antithesis rather than synthesis, passing in successive phases or stages to address collective social issues as well as to achieve individual self-realization. The Kharijis of Arabia, who were Sunnis, the Qarmatians and Assassins of Iran, Roshaniyya of India, not to say Shiah Islam in general, all passed successively from (*a*) the stage or phase of being protestant and puritan, holding up a mirror to the world, to (*b*) that of antinomianism, i.e. being branded as schismatics and heretics, and (*c*) came to end the cycle upon their conflict with and (usually) destruction by the authority of the sultan as well as the mullah—at any rate until the next heartbeat of discontent among the common people—or (*d*) by seceding and setting up in order to resolve the conflict a parallel religion and polity in some separate territory. This must be in outline surely the dialectic of religion and society that produced the legendary seventy-two sects of Islam foretold by the Prophet himself, as among the children of Israel, apart from the orthodox Sunni, the self-styled 'saved ones'.

Tariqat *and* hukumat

The Sufi leader and follower made socially conscious at some stage through this dialectic is also the Sufi armed, or at any event with both a secret *batini* doctrine and a clandestine subversive organization, and so destined to come into conflict with both the *shari'at* and the *hukumat*, which naturally combine together against his *tariqat* in this circumstance defined by the unity of religion and politics on both sides. It is the militant face, stage or phase of the movement of mysticism as against its pacific face of individualist esotericism, which is more usually represented in the literature as normal and permanent.

The Orientalist historian Hardy explains, for example, that 'extreme' Shiah sects, Ismaili and Qarmatian, had appeared early in upper Sind, and established a principality at Multan. Mahmud of Ghazni had defeated and dispersed them in 1005 and later, but they continued to exist underground though regularly 'slaughtered and imprisoned' for centuries.

> The Isma'ili and the Qarmatians owed their strength to the social discontent of later 'Abbasid times. . . . [It] made many humble people enemies of Sunni Islam. The latter appeared to condone the social ills of the time. On the eve of the Muslim conquest of India it was customary to stigmatize enemies as Qarmatians or Batinis.
>
> The Isma'ilis and the Qarmatians appealed primarily to the poor and the lowly, to peasants and artisans. The Qarmatians practised community of property, and according to their enemies, of wives also. They organized workers and artisans into guilds.
>
> The Isma'ili and Qarmatian [*a*] denial of the legitimacy of the sultanate, their egalitarian urges and their secret guild organizations caused the Delhi government as much alarm as [*b*] their rejection of the orthodox caliphate, schools of law, and theology scandalized the Sunni ulama (1958: 376, 383).

Let us retreat for safety next to the more moderate position of, for instance, the mystical orders of India in the thirteenth century, the Chishti and the Suhrawardi. The Chishti were spread all over north India and thoroughly oriented to God and the common people (*khalq*), albeit as individuals. The Suhrawardi, important in Sind and the Punjab, on the other hand, had royal patrons, raised no objection to the company of men of wealth and power in civil society, practised hereditary succession and lived well as family men, not redistributing immediately to all comers whatever had come to them unasked (*fatah*, charity). It is true that the Chishti leaders were also usually married men, but Nizami complains ruefully that no saint gave sufficient care to the upbringing of his children, and they uniformly neglected their households, with the result that most of the sons and family dependents of the Chishti saints turned out to be worldly people of civil society, 'unworthy of the tradition of the *silsilah*' (1961: 204).

In general terms, as already said, a mystic was expected to reject civil society and the world (*dunya*), which was often represented in the forms of 'treasure, woman or government service' (*shughl*). Leaders of the Chishti *silsilah* specially cut themselves off completely from the state, kings, politics and government in the early middle ages. A hierarchy of saints was established in north India, with a recognized chief saint and his accredited deputies (*khalifah*), and a network of *jama'at-khanah* (Chishti) or *khanqah* (Suhrawardi), according to territories or jurisdictions (*wilayat*). Nizami thinks that a careful analysis of the sites of the convents, hospices and retreats of the early Indo-Muslim mystics would reveal the fact that most of them were established outside the city and the society of privilege, 'in the midst of the lower sections of the Indian population'. Indeed,

> [t]hough within the political confines of the Sultanate of Delhi, the *jama'at-khanahs* of the early Chishti saints of India did not form a part of the Delhi empire. They were a world of their own (1961: 175, 237, 240, 261f).

The social effects of this somewhat 'ambivalent' position, therefore, happily combined both status or virtue and power—which *shari'at* and *hukumat* or Baghdad and Bukhara had not—but surely only in relation essentially to the individual and the esoteric. It is true that,

> when Shaikh Mu'inuddin Chishti was asked about the highest form of devotion, he replied that it was nothing but helping the poor, the distressed and the downtrodden.

And also that, in relation to the civil society:

There is hardly any social or moral crime against which the contemporary mystics did not raise their voice—slavery, hoarding, black-marketing, profiteering, wine and venery.

In fact, all kinds and classes of people, rural and urban, sought 'peace of mind' in the Sufi conventicle, and many of them took up residence and sought initiation there for the purpose.

[*a*] The [*zahiri*, externalist or exoteric] scholar shed away his intellectual arrogance and concentrated on the development and culture of his emotions; [*b*] the government servant developed hatred towards worldly power and authority and adopted service of humanity as his ideal in life; [*c*] the businessman dispensed with all his material goods and sought spiritual satisfaction in living a life of penitence and poverty (1961: 210, 236, 264).

It will perhaps appear churlish to point out that one has everything here for personal salvation, especially individual healing, spiritual training and initiation, not to mention the working of miracles (*karamat*, charismata), but nothing at all for a restructured social organization, labour or science.

There was always some degree of symmetry in India of course between *tariqat* and *hukumat*, the aristocracy of wealth and the aristocracy of poverty, the miracles of power and the power of miracles. It has been noted by Nizami that people reverentially often used political terminology to indicate the saint's position and sphere of spiritual activity. Thus Hansi, Hissar district in the Punjab, was customarily referred to as the 'frontier' between the Chishti and the Suhrawardi jurisdictions. Again,

The custom of prostrating before the king was prevalent both in India and Persia. [*a*] The Muslim rulers, both caliphs and sultans, adopted it very early. The Seljuqs, the Samanids and the Ghaznavids introduced it in their courts. If it reached the Delhi court through Persian channels, [*b*] Indian traditions brought it to the medieval *khanqahs.* All visitors to Shaikh Fariduddin Ganj-i Shakar and Shaikh Nizamuddin Auliya showed their respect to the great saints by laying their foreheads on the ground.

If it went too far, however, the conscience of *shari'at* would right the symmetry of *hukumat* and *tariqat*, so to say, at any rate in normal historical circumstances. Thus Shaikh Nasiruddin Chiragh (d. 1356), next in the line of succession to the Chishti *gaddi* of

Delhi, discontinued the custom, declaring it not lawful in Islam, in its aspect of *fiqh* or *shari'at*, to place one's head on the ground before a creature, but only the creator (1961: 94n, 176).

We may now take leave of the matter of the Muslim culture of medieval India, as we began, with the words of Habib of Aligarh as a summary.

Many Companions of the Prophet kept away from the empire which Amir Mu'awiya had formed because it was a thing of sin, an organization of the governing class based on the exploitation of the governed, while they were not in a position to oppose it. Most religious leaders, mystic [Sufi] and non-mystic [mullah], of the middle ages adopted the same attitude. The great Chishti Shaikhs up to the successors of Shaikh Nizamuddin Aulia (*circa* 1350 AD) followed this principle. After that, mystics of repute began to seek for their disciples among high officers or the kings of minor dynasties, who could afford to maintain them; at the same time they appointed their sons as their chief successors in order to maintain the prosperity of their families. Since these mystics lived under administrative protection, they lost touch with the [common] public [or civil society]. Muslim mystic thought, properly so-called, ceased to advance. A new type of Indian teachers appeared, who are best represented by Kabir (1961: x).

Habib goes on to remind us of Tara Chand's seventeen sects of religion that taught doctrines like Kabir's, held Hinduism and Islam to be equally true, and sought a correspondence, if not an equivalence, of Allah : Rama (divinity) :: guru : pir (*tariqat*) :: Hindu : Turk (*hukumat*) :: pandit : shaikh or mullah (*shari'at*). Like the Chishti order or sect before them, they were completely pacific and non-violent, appealed to the lower orders of society, cultivators, artisans and small capital traders, and did not seek the patronage of any sultan, raja, amir or vizier (1961: xi). Habib concludes that, even though small in numbers, they had acquired wide and deep influence, out of which a new India was to be born, and which I conclude was destined to try and produce a modern plural society in the permanent Indian sense and logic of unity in variety.

If anyone asks, nevertheless, for an explicit statement of my conclusion in terms of the mutual relation of the Hindu culture and the Muslim culture, a topic that we have been careful to

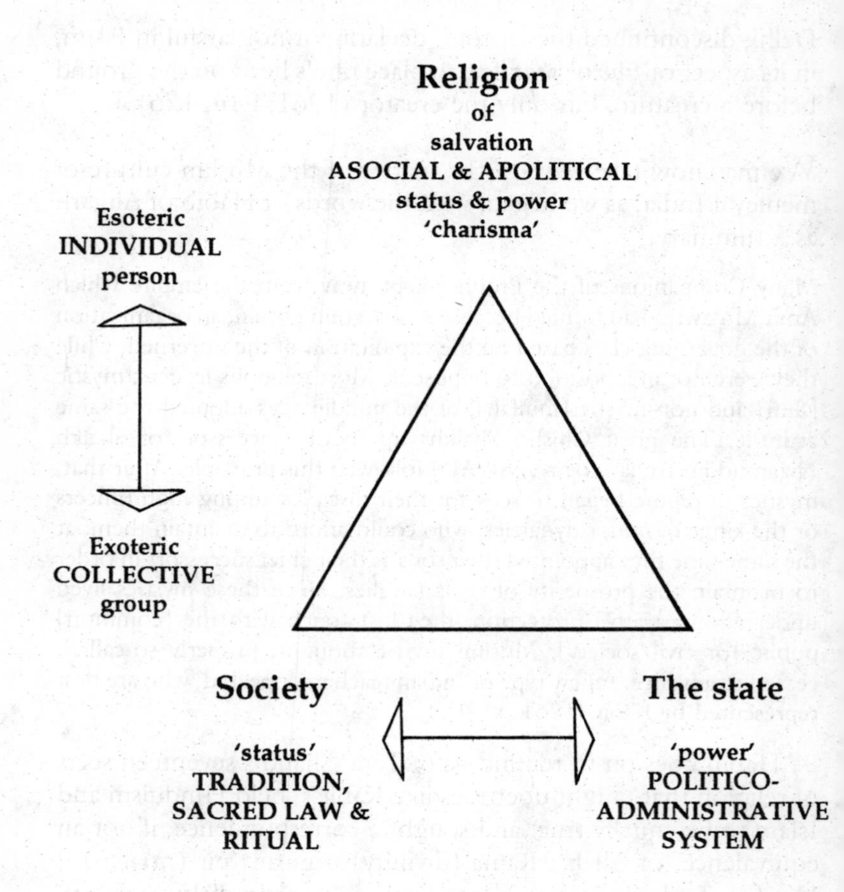

FIG. 3 Elementary Structure of Medievalism

avoid, one must confess that I have not here investigated how the two traditions understood each other, or failed to do so, but have advanced an hypothesis that this is the elementary grammar or syntax of how each of them understood itself and the two together themselves in the middle ages (fig. 3). I believe that the underlying structure, principles or logic and language of self-identity for the two cultures of India, Hindu and Muslim, was one and the same, a phenomenon analogous to the modern relation of the Hindi and Urdu languages which are surely different in lexicon, classical allegiance and script, but are one and the same in the syntax of intelligibility or meaning and effect, namely Hindustani, and which Gandhiji also wanted to be recognized as the national language of modern and free India.

3

Sikhism and Islam: Philosophy of Religion

Make kindness thy mosque, sincerity thy prayer-carpet,
what is just and lawful thy Quran,
Modesty thy circumcision, civility thy fasting, so shalt
thou be [called] a Musalman;
Make right conduct thy Kaaba, truth thy spiritual guide,
good works thy creed and thy prayer,
The will of God thy rosary, and God will preserve thine
honour, O Nanak.

Transl. Macauliffe 1909: I, 38

The Problem of Sikhism and Islam

It is apparent to everyone that Sikhism and Islam are two separate and different religions in theology as well as history, and there is no gainsaying the fact that the Sikhs are not Muslims, and the Muslims are not Sikhs, in any ordinary sense. Yet one finds that there is perhaps some mysterious affinity between the two systems of faith and morals in theory and practice, as there is in the relation between the followers of the two religions in south Asian history and politics. We shall attempt here a structural or semiological definition of the problem, which is even more intractable than the relation of Sikhism and Hinduism, and offer the outline of a solution in terms of simultaneity as well as succession, i.e. in terms of both comparative philosophy (synchrony) and comparative history (diachrony), with civil society as the silent third term of their mediation. By the civil society I shall implicitly refer, not so much to the commonsense society of classes, strata and groups and their relations, but to that underlying configuration of India, an aspect of the unity of summation, which mediates the interrelations of various social systems, modes of thought and codes of conduct and their reciprocal functions, and in particular the relation of religion and politics.

A modern Western scholar of the nineteenth century, Pincott, apparently relying on the original Punjabi sources, had concluded for himself that Sikhism might be treated almost as a sect or denomination of Islam. Pincott writes that the name Muslim or Muhammadan, in the various countries in which it exists, was allowed to cover differences in religious belief quite as great as those found between the views of Guru Nanak and of the Prophet Muhammad, and I shall refer to his problematic and conclusion later (1895: 594).

From a Muslim point of view, i.e. the insider's view, a sympathetic modern scholar might, more likely than Pincott, assimilate the Sikhs (*a*) to the status of *ahl-i kitab*, people of a book, a scripture or a revelation in time prior to the Quran; or perhaps more properly (*b*) to the status of the *ahl-i tasawwuf*, the Sufi tradition. Until very recently, in Afghanistan and central Asia or wherever society was under the regime of orthodox Sunni Islam, the Sikhs were classed with the Hindus (*c*) as *zimmis*, protected people of the covenant or obligation, entitled to tolerance but subject to certain discriminations according to the Hanafi law of the *shari'at*, whether or not they were liable to the poll tax (*jizyah*). The quality of Muslim responses to Sikhism found in modern history and civil society thus varies from these three categories of a spiritual, statutory and legal toleration up to or down to (*d*) the category of complete mutual hostility and conflict, in the sense of Muslims regarding the Sikhs as unbelievers and infidels (*kafir*), and so the object of *jihad* or an holy war, ultimately having only the choice of Islam or the sword.

One may say that, nevertheless, the spirit of the reformist Wahhabis of Saudi Arabia in west Asia, for example, must find some similarity with the essentially unitarian spirit, the anti-ritualism, anti-idolatry and the general Puritanism of Sikhism, if not also with its positively mystical side. Paradoxically, we learn that, in spite of this affinity or perhaps because of it, Sayyid Ahmad of Rai Bareli, the Wahhabi or Wahhabi-like leader of India, had raised an insurrection on the Peshawar frontier, declaring an holy war (*jihad*) against the Sikhs of the Punjab in AD 1826. The Sikhs replied in kind and the war lasted for several years, with the British turning a blind eye from their dominions, but I think that somehow it was unsuccessful as a *jihad* for either side. There are also several other such conflicts in history, which, if they are selected and strung together, will seem to support the conclusion that a permanent and undying political state of feud always existed between Sikhism and Islam: this is the problem of religious or cultural similarity and political conflict that has been set for us to redefine in more dialectical terms.

The general outline of Sikhism and its several distinctive features, to begin with, are best summarized in the words of a non-Sikh historian, Niharranjan Ray, which I simply quote in full as five points.

[*a*] To be able to achieve the integration of temporal and spiritual seems to me to have been the most significant contribution of Guru Nanak to the totality of the Indian way of life of medieval India . . . [where] one finds that, by and large, in thought as well as in practice, the temporal and the material were set in opposition to the eternal or perennial and the spiritual.

[*b*] . . . This sharp social consciousness characterized many of the Gurus, especially Guru Arjun and Guru Gobind Singh. Their concern for the lowliest and the lost, the human appeal of their religious aspirations, their regard for honest manual labour for earning one's livelihood and their intense dislike of a parasitic existence, their unceasing and prayerful concern for a clean and honest life marked by fearlessness, on the one hand, and their protestant attitude towards all kinds of sham and shibboleth in religion, society and politics, on the other, marked their *sishyas* out as a community distinct from the two other major communities, the Hindus and the Muslims. . . . The Sikh Gurus [also] took consciously a series of steps directed towards marking themselves and their followers out as a community with an identity of their own, clearly distinct from both the Hindus and the Muslims.

[*c*] Guru Nanak and his successors offered a new message and a new mission, both simple, direct and straightforward. The message consisted

in the recognition and acceptance of one and only one God in place of hundreds of gods and goddesses. He also told them that this God could be reached not through the intermediary of priests but by one's own honest efforts, through love and devotion and through God's grace, but following a rigorous course of discipline. He gave them a prayer and a routine as the keys to this disciplined way of life, and the life of a householder given to practical activity in matters of the world as much as in the matters of the spirit. The mission consisted in rejecting all external forms and practices of religions and spiritual exercises, meaningless rites and rituals, base and degrading social abuses and practices. Positively, it also consisted in the acceptance of the dignity of manual labour, and the social duty of making no distinction between the rich and the poor and of fighting the forces of evil. Here was a simple and straightforward message and mission, easy to understand and worthwhile following in practice and holding up as an ideal.

[*d*] . . . But what held these countless number of people together was neither the message by itself nor the mission by itself, not even by the two operating together. It was, to my mind, the institutionalization of both, and the organization that was built up stage by stage by the Gurus. . . . The [personal] leadership and the charisma of the Gurus served only as incentives and gave the necessary inspiration and guidance (1970: 70, 86, 96).

In conclusion, he also gives us a hint as to the Sikh theory of martyrdom, although it is not followed up.

[*e*] For the first time, fear of death, the darkest and greatest of all fears, was taken out of man, death not merely in the heat and tumult of war, but death in silent defiance of the most painful and tortuous tyranny. . . . Yet the universal God was ever their sole inspiration and ideal of social and individual activity (Niharranjan Ray 1967: 9).

Theology of the First Guru

It is popularly held that Guru Nanak (1469–1539), the founder of Sikhism and first guru of the Sikhs, received the call and declared his ministry at the beginning of the sixteenth century with the words or formula of negation, 'There is no Hindu, and there is no Musalman.' The common people who heard this emphatic and repeated affirmation in AD 1499 or 1507 on the north-west frontier of Hindustan went and reported the event to the khan, his former employer in service at Sultanpur, who called him and said, 'O Nanak, it is a misfortune that a steward such as thou shouldst become a faqir.' The qazi, or the legist of the *shari'at*, who was with the khan at the time, thereupon asked the new faqir to explain his affirmation of negation, to which Guru Nanak gave the reply, interposing his ethics between theology and sociology, 'To be called a Musalman is difficult; when one (becomes it) then he may be called a Musalman. . . . A (real) Musalman clears away self; (he possesses) sincerity, patience, purity of speech: (what is) erect he does not annoy: (what) lies (dead) he does not eat. O Nanak! that Musalman goes to heaven.' Then the local people, Hindus and Muslims, began to say to the khan that God himself (*khuda*) was speaking in Nanak (India Office London MS 1728, *Janam-sakhi*, transl. Pincott 1895: 585).

After this declaration of his ministry of God, say in 1500, there commenced the long period of his itinerant travels and exchanges with various religious figures, Muslim and Hindu, in the four cardinal directions of India and beyond as we shall see, before he finally returned to establish the faith and the community of Sikhism in his new place of domicile (Kartarpur, 'abode of God'). In the modern popular iconography of lithographed bazar calendars, Guru Nanak was until recent years represented as sitting in a sylvan natural scene with two attendants, Mardana the Muslim on his right and Bala the Hindu on his left side. He is always represented as sitting under a tree, but I think not in conformity with any known *asana* (Hindu or Buddhist); and he is shown attired in the headgear, not the turban, and the clothes of a Sufi or a faqir, but again avowedly not of any known religious order (Muslim).

Baba Nanak, as he is called with respect and affection by both Sikhs and non-Sikhs, was himself neither a Muslim nor a Hindu, and yet somehow he and his message were accessible to both groups at once, as a common language. Guru Nanak's office in the world was thus not so much to announce a third or a middle way, as is often wrongly supposed by the learned, but to seek to restore for all, Hindu as well as Muslim, the truth and the reality of the unity of God as the object of worship and the foundation of faith and morals. The theology of Sikhism is unitarian, in the dialectical sense of non-dualism, and it is not to be regarded as a variety of syncretism of monism, pantheism and theism, etc. The folk wisdom of India has surely got the message right, e.g. in the Urdu and Hindi saying about his mission and reputation that:

> *Nanak Shah faqir;*
> *Hindu ka guru,*
> *Musalman ka pir.*

This common Hindustani saying was appropriately quoted by Zakir Husain, President of India, at the time of Guru Nanak's birth quincentenary celebration (1969: v).

The original theological relation of Sikhism and Islam can be illustrated very well by the historical report of Guru Nanak's two dialogues with Shaikh Farid of Pak-Pattan, formerly known as Ajodhan and situated midway between Lahore and Multan. The original Fariduddin Shakarganj (1175–1265), who flourished at the turn of the twelfth/thirteenth century AD, was perhaps the most famous Muslim saint of the Punjab, widely adored to this day by Hindus as well as Muslims, including specially the Shiahs. He belonged to the well-known Chishti order of the Sufis, a branch of which was developed by one of his pupils, Nizamuddin Auliya of Delhi, whose prayers are believed to have saved the city from the Mongol visitation in 1303. In the historical interviews with Guru Nanak, therefore, we must suppose the name Farid to be the *nom de plume* of his contemporary successor at the shrine of Pak-Pattan. The holy Sikh scripture, the Adi Granth, has included in it some 112 verses attributed to Baba Farid, and the Sikh principality of Faridkot was so renamed in his honour from the original Mokalhar.

It is recorded that, at the first mutual greeting of the two, Shaikh Farid exclaimed, 'Allah, Allah O Darvesh!' and to this Guru Nanak replied, 'Allah is the object of my efforts [*jahad*, same etymology as *jihad*], O Farid! Allah, Allah (only) is ever my object' (Pincott 1895: 587).

Abdul Majid Khan, who was both a Muslim and a Gandhian, writes (*a*) that in one of the old *Janam-sakhis* of Bhai Bala, who is only a supposititious narrator, Guru Nanak is stated to have uttered the verse, *'Ditha nur Muhammadi, ditha nabi rasul;/ Nanak qudrat dekh ke, khudi gai sab bhul'*, which translated means that he saw the light of Muhammad, the Prophet and the messenger, he saw the glory of God and his creation, and he forgot his egoism or self (1967a: 227). There is nothing that is incredible about the story, although I cannot say that it is well-established. The 'light of Muhammad' is here not a metaphor but a technical philosophical term, in Arabic *al-haqiqat al-Muhammadiyah* or *haqiqat al-haqa'iq*, the original essence of the Prophet, which is believed to have been created before all things from the light of God. It is the creative, rational and animating principle of the universe, and when the Almighty resolved to make the world, he divided this light into four parts, the fourth of which was again sub-divided into four parts, and so on. From the first part he created the Pen, from the second part the Tablet, etc., until we come to the 'remaining portion of creation', the world, which was apparently made of only a sixty-fourth part (see Hughes 1895: 162).

So far as the Adi Granth is concerned the doctrine of light and its role is sufficiently clear, and it is included as stated in the opening verse of Kabir, 1440–1518, *Vibhas-prabhati*: 'In the beginning, God created the light; all are creatures of his creation (*qudrat*)./From that one light, the whole cosmos came into being, (including) who are good and who are evil.'

It is one of the heterodox, mystical, Neoplatonist or Gnostic ideas, along with (*b*) that of the Prophet as the perfect cosmic man (*insan-i kamil*), the microcosm, which are associated with the name and work of Ibn Arabi, the Spanish mystic (*fl.* 1200), at the point of maturation of Sufism as an institution, and which found a response in Shiah Iran and in medieval India. The author of the verse, whether it was Guru Nanak or some other Punjabi, would have thought it all of a piece with the possible, although not the common, Muslim view (*c*) that all truth from God revealed anywhere at any time is Islam.

Guru Nanak's words during his second dialogue on the unity of the godhead with Shaikh Ibrahim of Pak-Pattan (d. 1552), the original Shaikh Farid's successor in that Chishti *gaddi* and known as Farid Sani or Farid II, express the same theology (transl. Pincott 1895: 587).

Thou thyself (art) the wooden tablet;
Thou (art) the pen;
Thou (art) also the writing upon (it).
O Nanak, why should the One be called
a second?

In other words, the founder of Sikhism sought to draw a clear line of distinction between the religion of Islam, which he would accept, and what might be called Muhammadanism, which he would reject. Shaikh Ibrahim then asked the guru for a further historical explanation, moving from the theology to the sociology of the question.

Thou sayest, "There is One, why a second?"
but there is one Lord and two traditions.
Which shall I accept, and which reject?
Thou sayest, "The only One, he alone is one";
but the Hindus are saying that in (their) faith
there is certainty;
and the Musalmans are saying that only in (their) faith
is there certainty.
Tell me, in which of them is the truth, and in
which is there falsity?

Guru Nanak replies to the question simply, directly and firmly:

There is only one Lord, and only one tradition.

The same mystical theology of the first guru is expressed in the first verse of his *Japji*, the opening, as translated by Macauliffe (1909: I, 195–6).

The True One was in the beginning, the True One was in the primal age;
The True One is now also, O Nanak; the True One also shall be.
. . . By His order bodies are produced; His order cannot be described.
By His order souls are infused into them; by His order greatness is obtained.
. . . All are subject to His order; none is exempt from it.
He who understandeth God's order, O Nanak, is never guilty of egoism.

There is a great deal more that might be quoted from the

sources in a similar vein, but it will suffice here to conclude in anticipation. (*a*) Sikhism is to be simply defined as the religion of the followers in this world of the One tradition and the tradition of the One, the path of the disciple and the community of saints, the discipline of the name and self-sacrifice as the condition of God's grace in a society for salvation. (*b*) The followers of Sikhism came largely from among the Hindus, and only a few from among the Muslims, but it is noteworthy that the figure of Baba Nanak is the one and only non-Muslim figure, not excluding the figure of Mahatma Gandhi, to hold a steady interest among the Muslims of India up to the present day. One need not agree with the view (*c*) that his pious object of renewal on the twin bases of unitarianism and self-abnegation was later on defeated by political causes and by the warlike nature of the people of the Punjab. One may agree, however, (*d*) that, at any rate in its historical and cultural elements and without denying its uniqueness or originality, the religion of Guru Nanak was 'based on Hinduism, modified by Buddhism and stirred into new life by Sufism' (Pincott 1895: 591ff).

The best authorities are agreed that the founder of Sikhism held that the soul of man was a ray of light emanating from the light of God, so that in its original and natural state the human soul was sinless and free. He further identified human egotism (*haumai*) as the source of evil, and the sense of duality as a snare and a delusion (*maya*). God is personal, the secret that dwells in each person's heart, his word (*sabad*) is embodied in the guru's word (*bani*), and he is also impersonal and beyond all comprehension. God is both within and without his creation: he is transcendent (*sa[r]gun*) as well as immanent (*nirgun*). In the scripture, the two most frequent appellations of God are the true creator (*sat-kartar*) and the true name (*sat-nam*). The name of God is the form of God, and the form of God is the name of God; of all modes of worship, the worship of the name is the best.

God reveals Himself; indeed, He is [*a*] the *Sabad* or the Word; He is [*b*] the *Nam* or the Name, [*c*] the *Guru*, [*d*] the *Hukam* or the Divine Order, [*e*] the *Sach*, the Truth. It is these and Unity that one must concentrate and meditate upon . . . by loving devotion for and adoration of God, and by endless repetition and remembering of His Name (*Nam-simaran*) (Niharranjan Ray 1975: 69).

The affinity of this theology, specially the theory of the name, with the Sufi tradition (Muslim) as well as the Sant tradition (Hindu) is already sufficiently well attested. It is sustained up to the disciple's goal of abolishing, under the guidance and the mediation of the guru, the duality between the self of man and the effulgence of divinity: to lose the self entirely is to become divine, to assert the self contrariwise is to deny God.

At the same time, one must add that Guru Nanak strongly departed from the majority of Sufi and other Hindu religious orders in his environment, in the sense that he prescribed the life of a householder and working man or woman as necessary for salvation, as against the life of renunciation of the world or of monastic asceticism. 'I take shelter in nothing save the guru's word and the congregation (of saints)' (dialogue with the *siddhas*). Guru Nanak himself returned, it is believed, to the married state and the working life of agriculture, after the period of his travels as a mendicant, itinerant preacher and missionary. He eventually also passed over in the succession to guruship his son, who had renounced the world and who became the origin of the Udasis, a separate sect.

'Without the practise of virtue there can be no worship' (*Japji* 21); and elsewhere, 'Truth is the greatest of all, but greater (still) is truthful living.' In the Adi Granth generally this world is called one's father's house and the next world one's father-in-law's house. Adopting the point of view of the woman, the desired relation of the two worlds or of man to God is assimilated to the relation of the bride to her spouse, in the common fashion of many Sufis and Sants.

It has been further suggested by an heterodox Muslim opinion (Qadiani) that the *Japji*, the chief composition of the first guru, is in some way his commentary on the Quran, but we have no independent confirmation of this. Many Sikhs do believe, however, that Guru Nanak composed and recited the *Japji* during his visit to the city of Baghdad in west Asia, *c.* 1520. All Sikhs consider the *Japji* to be the epitome of Sikhism, the key to the scripture, and every Sikh should learn it by heart for the morning divine service. Incidentally, the *Japji* is unique in the scripture in that it is the only composition in the Adi Granth that, like the Quran, is not set to any musical *raga*. All the other compositions are set and arranged according to one or other of the thirty-one *ragas* to which they were composed.

The Life of the First Guru, 1469–1539

As represented by the culture of popular tradition, the lives of the gurus, which stand next after the scripture and the non-scriptural writings of the gurus in authority, are most often given in the form of a travelogue and dialogues in relation to the religious question, specially the *Janam-sakhis* or life histories of Guru Nanak, where the supposititious narrator is Bala the Hindu, but which are in fact a collective and anonymous product. After describing the circumstances of his birth and childhood, and following the moment of his assumption of the ministry of God, up to which point one may simply see the semiotics usual to the gurus of Hindu sects of the period, the most frequent interlocutors of Guru Nanak are represented, I think, by (*a*) the khan, the governor or the ruler, (*b*) the qazi or the mullah and (*c*) the pir or the shaikh, or on occasion their Hindu equivalents. These three figures may be taken to represent his new attitude towards the medieval institutions of (*a*) ***hukumat*** or the rule of the state power, (*b*) ***shari'at*** or the law of ritual and society, and (*c*) ***tariqat*** or the inner religion of the other world.

The new faith and way of life was to be defined, propagated and instituted as a new structure of relations among them, explained as the way of Indian modernity, unity in variety, not only in the relations among different traditions, but among the different aspects of human life, including the socio-political and the economic. I will now briefly illustrate this structure below from the single composite popular life history usefully provided in English translation by the labour of Macauliffe (1909: I, 34ff, 40ff, 78, 84, 102ff, 121ff, 176, 281).

(*a*) First of all, Guru Nanak receives the divine call and his commission, having mysteriously disappeared in the water for three days after bathing in the forest. This is the moment when:

God said to him, "I am with thee. I have made thee happy, and also those who shall take thy name. Go and repeat Mine, and cause others to do likewise. Abide uncontaminated by the world. Practise the repetition of My name, charity, ablutions, worship and meditation. I have given thee this cup of nectar [of immortality], a pledge of My regard." The Guru stood up and made a prostration. . . . Hereupon a voice was heard, "O Nanak, thou hast seen My sovereignty."

(*b*) Next, when he was about to declare himself and his mission, with the affirmation of negation, 'There is no Hindu, and there is no Musalman', a mullah is brought in to try and reason with him or to exorcise his apparent madness of taking non-dualism to the extreme. So Guru Nanak complains to the world about himself:

Some say poor Nanak is a sprite,
some say that he is a demon,
Others again that he is a man. . . .
When [a] man loveth the Lord and deemeth himself worthless,
And the rest of the world good,
he is called mad.

(*c*) Along with this surely went Guru Nanak's social message from the beginning, e.g. when he told Malik Bhago that the bread of the rich, which he always refused, was full of the blood of the exploited poor. Mardana the Dom or minstrel, a Muslim who died in 1522, was his attendant on almost all his travels, and used to accompany his compositions on the *rabab* or rebeck, a stringed musical instrument of probably Arabian origin.

The Guru, in company with Mardana, proceeded to Saiyidpur, the present city of Eminabad, . . . [to] the house of Lalo, a carpenter. . . . Malik Bhago, steward of the Pathan who owned [the village of] Saiyidpur, gave a great feast, to which Hindus of all four castes were invited.

The guru, who was the son of a Khatri (derivative of Kshatriya?), replied, 'I belong not to any of the four castes; why am I invited?' But the wealthy Malik Bhago was not to be appeased and demanded an explanation of his absence.

Upon this the Guru asked Malik Bhago for his share, and at the same time requested Lalo to bring him bread from his house. When both viands arrived, the Guru took Lalo's coarse bread in his right hand and

Malik Bhago's dainty bread in his left, and squeezed them both. It is said that from Lalo's bread there issued milk, and from Malik Bhago's, blood. . . . Lalo's bread had been obtained by honest labour and was pure, while Malik Bhago's had been obtained by bribery and oppression and was therefore impure.

'If clothes become [ritually] defiled by blood falling on them,' the guru said similarly to the qazi who was dutifully carefully engaged in beheading a goat, 'how can the hearts of those who drink human blood be pure?'

(*d*) There is the moment of temptation by Kaljug, the Satan who came to tempt the guru in the wilderness beyond Kamrup (in Assam?), promising him the wealth of the world, the secret esoteric power of working miracles and the sovereignty of the East and the West, if he would but abandon his mission. What would he do, replies Guru Nanak, with what Kaljug offered him, which moreover belonged by right to others?

Incidentally, I have quoted the travels of Guru Nanak to the east and the west, but have omitted here his travels to the north and the south, the last evidently referring to a Buddhist king of Sri Lanka and his discourse there (see Niharranjan Ray 1970: 6n).

(*e*) The qazi and the shaikh or the pir repeatedly advise Guru Nanak not to 'put his feet in two boats'. They ask him to choose to lead either the secular or the religious life, and not to try and combine them, to try and seek either for high position or for God, but the guru remains steadfast in his resolve.

> Put thy feet on two boats and thy property also on them:
> One boat may sink, but the other shall cross over
> [the body may perish, but the soul shall be saved].
> For me there is no water, no boat, no wreck and no loss.
> Nanak, the true One is my property and wealth,
> and He is naturally everywhere contained.

(*f*) The shaikh, here again referring probably to Shaikh Ibrahim a successor of Farid in his Sufi order, then positively asks of his way of life, 'Give me such a knife as will make those who are killed with it acceptable to God.' The guru replies,

> Truth is the knife, truth is pure steel;
> Its fashion is altogether incomparable.

Put it on the hone of the Word,
And fit it into the scabbard of merit. . . .
If man be slaughtered with it,
he shall go to meet God,
O Nanak, and be absorbed in the sight
of Him.

Finally, (*g*) Guru Nanak has to give his answer to the invitation of voluntary conversion to Islam offered as a challenge. The established Islamic credal principle is that revelation could not be received and worship or adoration could not be performed without two beings, God and the Prophet, so that the second would be the intercessor for man. For example, Miyan Mitha of Mithankot, Dera Ghazi Khan on the north-west frontier, challenges him, saying,

The first name is that of God,
the second that of the Prophet.
O Nanak, if thou repeat the Creed,
thou shalt find acceptance
in God's court.

The guru replies in accordance with the theological principle stated in the Sikh scripture and elsewhere that revelation or the word itself is the guru and the guru is the word or revelation. 'Why should we worship a second who is born and dieth?'

The first name is that of God;
how many prophets are at His gate!
O Shaikh, form good intentions,
and thou shall find acceptance
in God's court.

It remains the principle of Sikhism as later confessed and stated before the emperor as well as the people, according to the Sikh chronicles by Guru Arjun, the fifth guru and the first martyr, and whether one likes to interpret it exoterically or esoterically, that

The unseen is true and without anxiety.
Many Muhammads stand in His court;

with the equal addition in relation to the Hindu trinity of 'Brahma, Vishnu, Mahesh', the god(s) of creation, sustenance or preservation and destruction, that

At God's gate there dwell . . .
thousands of Brahmas, of Vishnus and
of Shivas.

In other words, just as he would willingly have the faith of Islam as unitarianism and submission to God, but not as reduced to Muhammadanism, if that should be at all possible, so he would likewise have the God of Hinduism without Brahmanism or *varnashramdharma*, as I have argued, thus leaving him to uphold and embody under a new configuration *sanatanadharma*, the eternal religion, in his time and place.

(*h*) The first guru of the Sikhs, when he purportedly made the journey as a pilgrim to Mecca, the holy city of Islam, ultimately tells the qazi and the mullah that, without knowledge of the one God, humility and good actions, neither the Muslims nor the Hindus should obtain entrance into the court of God, the true king: Satan had led them both astray, as he says, 'along his own flowery way'.

As to himself, this was his summary reply in the dialogue with the qazi of Sultanpur, who had asked Guru Nanak who he was, and what he thought he was doing:

Since I am nobody, what shall I say; since I
am nothing, what can I be?
As God made me, I act; as He told me, I speak;
I am thoroughly defiled with sin, and
desire to wash it away.
Though I know nothing myself, yet I teach others;
such a guide am I.

The Tenth Guru

For its first two hundred years during the sixteenth and the seventeenth centuries of the Christian era, i.e. from the time of its founder and first guru, Guru Nanak, up to the time of the tenth guru, Guru Gobind Singh (d. 1708), the commonwealth, confraternity and congregation of Sikhism was under the direct rule and guidance of a succession of mortal, human and personal gurus, the light of each of them being lit 'as one lamp is lit from another', making the succession to guruship also simultaneously impersonal, indivisible and continuous. After that time, or more accurately after the day of the institution of the Khalsa at Anandpur by Guru Gobind Singh on the Punjabi new year's day of Baisakhi in AD 1699, the scripture itself was to be taken as the eternal and non-human universal guide, the Adi Granth when installed in the centre of the temple, duly opened and read aloud in the congregation as the Guru Granth Sahib.

The fifth guru of the Sikhs, Guru Arjun, had himself edited, collated and reduced the scripture to writing in the vernacular at Amritsar, the city of the golden temple and its pool of immortality, in 1604. Soon thereafter he also became, for public refusal to deny his faith, the first martyr of Sikhism, tortured at the hands of the imperial authority at Lahore. Guru Arjun is, therefore, the keystone of the arch of Sikhism, of which the first guru and the tenth guru are the two pillars, and we shall have more to say about him in the next chapter. His symbol in the eyes of history is that of the city of Amritsar, rather than that of imperial Lahore where his shrine remains, if that of the first guru is Sultanpur or Kartarpur in the west and that of the tenth guru is Anandpur in the east.

On that new year's day, Guru Gobind Singh, completely laying aside all traditions of caste, occupation and lineage, ***karma*** as well as ***dharma***, first initiated by a new form of ceremony, which we have discussed in the first chapter, the five loved ones or the five lovers, who had already, each of them, individually offered their heads to him in obedience to his call of the faith. Then, standing before the five with folded hands, he asked them as a collectivity to administer the same baptismal rite of the spirit and the sword of gnosis to him, saying, 'You are in my form and I am in yours.' The community of the Khalsa, as it was now called, was thus given its position as equal to that of the guru: 'the Khalsa is the guru, and the guru is the Khalsa.' The future guruship, its charisma and functions, already united in the leadership as 'impersonal, indivisible and continuous', were now transferred to the scripture-in-the-community. It was now established to be constituted of two component parts, the body and the word or the name, whilst its functions of interlinking new beginnings, mediation and revelation, remained as before. Guru Gobind Singh himself nominated the collectivity of the elect as his own body, 'I shall ever be among five Sikhs assembled'; and the Sikh scripture, the Guru Granth Sahib, as embodiment of the word or the name, 'Obey the Granth Sahib; it is the visible body of the Guru' (Macauliffe 1909: V, 93ff, 189, 243, 294).

The tenth guru of the Sikhs therefore became in effect the disciple of his disciples at the new revolutionary moment of reversal, inversion and reflection of the leader/follower relation, and thus conferred on the Khalsa, the community of the pure and the elect—to which all Sikhs may or may not belong, the sovereignty of both spiritual and temporal self-rule, freedom in society and responsibility under the guidance of the divine word or the holy scripture, with all the ten gurus as its fixed exemplars (Khalsa Raj). The Khalsa was to be, since its inception, a society for salvation and self-realization, unitarian in religion, vernacularist in culture and democratic in politics; this was in its nature, its constitution and its modernity of non-dualism. This historical event completed the threefold identification of the godhead as the guru, the word as the guru and the congregation of the disciples as the guru. Each of them is an identification of reciprocal embodiment and participation, so producing the archetypal Sikh trinity of Guru, Granth and Panth (the way). In India perhaps this recalled equally the Buddhist trinity of the Buddha, the Dhamma and the Sangha; and the threefold equation in Islam of the unity of the godhead (*tauhid*), the scripture as the word of God (*kalam-i Allah*) and the community of the faithful (*ummat*), but I would not press the point.

One must say something now about the purely theological attitude of the tenth guru, Guru Gobind Singh, toward the religion of Islam. It is often stated by historians and others that he was brought up under Hindu guidance and influence during his youth; that in search for the power of offence as well as defence he became a worshipper of the goddess Durga or the Devi in the occult tradition of Tantrism; and that he was the furthest removed from Islam, specially as compared to Guru Nanak, and indeed opposed to it independently of the political conflict with the Mughals that was to follow in his later life.

We may call as witness to the contrary the opinion of Mohan Singh Diwana, the doyen of Punjabi scholarship, who stated on the occasion of his birth tercentenary, that the tenth guru of the

Sikhs held as securely as his predecessors to belief in the one transcendent and immanent God. God, who rules the world according to his will and pattern (*hukam*), but yet wishes, calls and requires his creatures to adore and glorify him and his works, specially by being witnesses to his attributes and qualities as against his essence, which is inimitable. The underlying view in the tenth guru's writings therefore continues to be that the salvation of man lies in his life's resignation to the will of God or the divine order, and his winning grace through disciplined self-sacrifice and emulation of the qualities and attributes that the gurus expounded and all bore living witness to. Surely there is no foreignness to the Islamic mode(s) of thought in this theology, specially in the terrain of India, whatever else may be true of Islam in terms of its origin and sources.

Diwana has given as part of his evidence a selected list of what I would call three kinds of Arabic-Persian usages, key terms borrowed under Sikhism. (*a*) Terms relating to divinity as they are to be found in the writings of the tenth guru, which have been collected as the Dasam Granth separately from the original Sikh scripture, the Adi Granth, but some of which are included in a Sikh's daily devotions by common consent and tradition. In addition to these twenty-five examples of names, attributes and qualities of the godhead, which I shall list below in a moment, there were evidently two other classes of key terms that were taken over by the gurus of the Sikhs from the Arabic-Persian tradition of India. (*b*) Terms relating to the arts of warfare, kingship and the state, which are simply transposed into the Punjabi, but often with a shifted emphasis that Diwana has discussed. (*c*) Terms that were originally a part of the *shari'at* lexicon, relating to the laws of Muslim ritual and society, but which are used by the Sikh gurus to refer to human ethical and moral values of civil society and universal categorical imperatives, as in the example from Guru Nanak given at the head of this chapter (Quran, the creed, mosque, liturgical prayer, prayer-carpet, circumcision, fasting, spiritual guide, rosary, Kaaba in Mecca).

Whether the whole of Sikhism was thus 'assimilative as well as reactionary', as Diwana says in relation to the influence of Islam or whether it was chiefly original and an innovation, is a large question that I shall not attempt to decide. The answer will depend partly upon whether and to what extent the tradition of Islam itself can be detached, as I think, from the two cardinal points of Arab nationalism with which it is originally enmeshed or in which it is embodied in history and society. (*a*) The finality of the Prophet Muhammad's revelation and therefore of his mission; and (*b*) its absolute identity with the Arabic language, as against the vernacular, whatever that might be.

The latter point of contention can be restated here in the form that Sikhism always implicitly claimed to be the complete truly vernacular religion for modern times, and so wholly democratic in its method and praxis. This vernacular and democratic aspect of Sikhism seems to still give equal metaphysical offence to the classical and aristocratic elements in Hinduism as well as Islam, who wish their revelation and its ritual or liturgy to be exclusively in the sacred original Sanskrit or Arabic languages. For instance, within the historical Islam of India, it is known that Bayazid (b. Jullundur, AD 1525), Pir-i Roshan, 'apostle of light', an Ansari shaikh of a kind of Shiah and Ismaili or Sufi persuasion, was a missionary to the Afghans or Pathans, whose vernacular he had adopted or wanted to adopt for the liturgy. Bayazid Roshan and his family and followers, the Roshaniyya sect of the north-west frontier which I have already cited, suffered frequent persecution and martyrdom, later on to the point of extermination, for this kind of thing at the hands of the Mughals, including by the supposedly liberal emperor Akbar himself in person, upon the usual medieval charges of denying orthodoxy and practising 'communism' of property and women among themselves.

For the moment, however, let us merely set out Diwana's list of twenty-five in comparison with the ninety-nine names of God in the Arabic-Persian originals, with which there is some overlap. It should be apparent to anyone who makes the comparison that in the hands of Guru Gobind Singh we do not have the usual simple alternative of either cultural borrowing or its presumed opposite, political hostility, but something of both, a mediating third term of co-existence. The compilation in evidence, which is simply an example, shows both a scholarly and loving familiarity with the tradition of Islam, the religion of submission to God, as well as a resoluteness to follow one's own chosen way and vision of revelation. In no way can it be taken, I think, to support the conclusion of an historian writing in the *Cambridge history of Islam* that, departing from its early origin, 'Sikhism became the implacable adversary of Islam in north India' under the Mughals (Qureshi 1970b: 61).

The unfathomable (*'amiq*);
the friend (*rafiq*);
the moon of moons (*mah-i mahan*);
the forgiver of sins (*'afiv al-gunah*);
the aware (*kalim*);
the lover of the poor (*gharib al-parast*);
the master of the word (*sahib-i kalam*);
the light (*husan al-chiragh*);
the speech (*zaban*);
the conjunction of opposites (*sahib-i qiran*);
the dweller in paradise (*bahisht al-niwas*);
the time and space (*zamin al-zaman*);
the thought (*khiyal*);
the mystery (*'ajaib*);
the vanquisher of foes (*ghanim al-shikan*);
the dispeller of foes (*ghanim al-kharaj*);
the beauteous countenance (*husan al-wajuh*);
the time of time / death of death (*zaman al-zaman*);
the dear one (*al-'aziz*);
the lord of the lands (*sahib-i diyar*);
the just (*dadgar*);
the giver of the arts (*har hunar*);
the spring (*naubahar*);
the orderer of the universe (*jahan ra tui bastah-i band-o-bast*);
the religionless one (*amazhabe*).

The last term adduced by Diwana here is perhaps the most novel usage of all, derived from the Arabic word *mazhab*, school of religion or law, with the new addition of the Sanskrit prefix of negation. Yet we may apparently rest assured that Guru Gobind Singh means what he says and that he says what he means (transl. Mohan Singh Diwana 1967: 254).

Thou art, O God, beyond all religion;
Thou hast no religion, except it be Godliness;
Thou hast created one after another all the religious
systems and destroyed them.

An Indian Muslim View Today

On the occasion of Guru Nanak's birth quincentenary (1969), the paper of M. Mujeeb, vice-chancellor of Jamia Millia Islamia, New Delhi, had perhaps suggested a solution to the problem of Sikhism and Islam. Mujeeb first makes an inventory of what he sees as the common points between the two religious systems—although he does not take notice of the Sikh theory of martyrdom in relation to the Shiah tradition, perhaps because of its political aspect (see chapter 5).

The features which appear common to Islam and Guru Nanak's teaching are: [*a*] belief in one God, at once transcendent and immanent, the God of all mankind, [who is] not to be represented by any physical symbol; [*b*] the equality of all men as men; [*c*] the organic fusion of the spiritual and the worldly life, i.e. of worship with the fulfilment of social obligations; [*d*] a community life based on work, worship and the generous sharing of what is earned; [*e*] a *sadh sangat*, or organized community living as an expression of the religious ideal; and [*f*] *dhikr*, the repetition of God's name as a form of prayer.

Now Mujeeb unexpectedly does not go on to draw the usual inference from this list that Guru Nanak's spiritual function was to attempt a synthesis of the two systems that would claim to be a 'higher' form of religion, which being neither exclusively Muslim nor exclusively Hindu would be theologically acceptable to both of them. He says rightly that such an inference would be a form of rationalization that does not adequately reflect the truth and the reality of the guru's experience of revelation.

Guru Nanak's spiritual experience was a unifying force because it was genuine; it does not claim to be genuine because it was a unifying force. . . . He was the founder of a new faith, and not a religious reformer.

At first sight it might appear strange that a Muslim should argue against the view that Guru Nanak was influenced by Islam in the usual historical sense,

but the belief that *Din*, the true religion, has been revealed to all men [through a prophet or a scripture] is one of the basic doctrines of Islam; and this belief is confirmed, and not weakened, by the assertion that

Plate 1

Guru Nanak, the first guru, as commonly represented in a sylvan setting, attended by Bala the Hindu holding a fly-whisk on his left and Mardana the Muslim with a rebeck on his right. At the beginning of the modern period, *c.* 1500, his utterance or revelation of the new scripture (*guru-bani*) destined to constitute the new liturgy is in the vernacular, the speaking voice of popular culture, also often depicted as a green parrot sitting in the tree or in a cage. The guru's own form of dress, head-gear and posture of sitting are also novel, and do not belong or correspond to the known systems or traditions, Hindu or Muslim, and he speaks the renewed common language of non-dualism or 'unity in variety'.

Victoria and Albert Museum, London, IM–2 (125)–1917, c. 1875

Plate 2

The serene martyr, Guru Arjun, the fifth guru, was sentenced to death with torture (Pers. Mongol, *yasa*) in 1606 by judgment of the emperor, Jahangir, upon charges of blasphemy and high treason (in relation to succession to the throne). He is shown in his last dialogue with the qazi of the state, refusing to recant even a word of the scriptures handed down from above, witnessed, compiled and written by him at Amritsar, 1604. Hot sand is being poured upon his head while seated on a hot plate; this torture follows the way that popcorn is commonly made in India. Alternatively, his interlocutor is identified as Miyan Mir, faqir of Lahore, who has come to ask if he can be of help; Guru Arjun motions him to look heavenward. The faqir is vouchsafed a vision of hosts of avenging angels ready to come to Guru Arjun's aid, which he refuses, saying that he must use his

(*continued opposite*)

Plate 3

The heroic martyr, Baba Dip Singh, d. 1757, who had vowed to free the site of the Golden Temple at Amritsar, recently desecrated and occupied by General Jahan Khan (and later on destroyed by Ahmad Shah Abdali, the Afghan). Here he is shown as a greybeard in a losing battle leading his guerilla force against regulars, hopelessly outclassed and outnumbered; but he contuines to fight and advance towards the temple holding his own severed head in his left hand, wearing an expression of perfect composure (detail).

Author's collection, popular calendar art, Delhi, 1993

continued from Plate 2

continued from Plate 2

power to set an example of patience: 'The true test of faith is in the hour of suffering.' The young woman on the right, who bribed her way at night into the citadel of Lahore which was his prison, is the daughter of a pious Sikh and the daughter-in-law of the guru's betrayer, Chandu. The Sikh chronicles say that she went home and gave up her soul after hearing the final news of the guru's death by disappearance after bathing, attended by five Sikhs, in the river Ravi (Gk, *Hydraotes*).

Author's collection, popular calendar art, Delhi, 1993

Plate 4

Guru Gobind Singh, the tenth and last guru, meets face-to-face Guru Nanak, the first guru: a Pahari school miniature painting, Mandi, *c.* 1700–20, which is not unique as two very similar ones are now in New Delhi and Venice. The two aspects of Sikhism represented by the two gurus, the pacifist or quietist and the militant, attired as the prince and the faqir, are here shown in simultaneous co-existence (synchrony), and not as an historical evolution (diachrony). The essential tension of the two was truly united and/or resolved in the person, work, life and death of Guru Arjun, the serene non-violent martyr. It was perhaps falsely, or at any rate only secondarily, resolved in the person and the career of Banda Singh Bahadur, military leader and heroic martyr, d. 1716, who is probably the second royal soldier shown on the right. The second faqir shown with a rebeck on the right is Guru Nanak's usual Sufi attendant, Mardana.

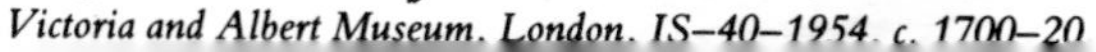

Victoria and Albert Museum, London, IS–40–1954, c. 1700–20

Guru Nanak's teachings present an independent and original spiritual experience. The Sikh is not obliged to the Muslim, nor the Muslim to the Sikh; and their faith in their own religions should be all the stronger because of any confirmation of the one by the other. If they walk steadfastly on what is their [own] true path they will discover that their paths and their goals are the same. That discovery itself will be a spiritual experience, an "occurrence of the heart", a fulfilment of what God, in the story related by Jalaluddin Rumi, said to Moses: "You have been sent to unite, and not to divide" (Mujeeb 1969: 114ff).

Evidently this view of Mujeeb is not unique, but has a tradition of Indian Islam behind it, whereby a third mediating category of *muwahhid*, monotheist or unitarian, had emerged from the usual Muslim/*kafir* opposition, the same Arabic term also being used later as a self-description by the Wahhabis of west Asia, who can scarcely be described as Sufi. India had thus added, in the period of Kabir and Guru Nanak, a fifth category, if you like, to the four non-Muslim categories recognized elsewhere and with which we began this chapter, namely, *ahl-i kitab, ahl-i tasawwuf, zimmi* or *kafir*.

S.A.A. Rizvi, who has made a study of the movement, writes that Shaikh Rizq Allah (1492–1581), when a mere boy, asked his father regarding Kabir, 'Is a *muwahhid* different from a *kafir* or a Muslim?' Guru Nanak belonged to this category, therefore, whose spiritual experience and vision went beyond the Hindu and Muslim idioms of orthodoxy, and he was recognized as a *muwahhid* in the common language. 'He is at once a Sufi and a *muwahhid* Sant'; and Rizvi thinks that the *Janam-sakhis* seek to present Guru Nanak as a 'perfect man' in the Sufic sense of Ibn Arabi. He did not turn pantheist, however, as some Sufis tended to become, but adheres remarkably to his version of monotheism, in the sense of devotion to the Supreme Being alone and not to any of his incarnations (Rizvi 1975: 200, 203, 211ff).

Indeed, the same unitarian qualities form the basis of the praise of the guru in the present century by, of all people, Muhammad Iqbal, philosopher of Pakistan, who refers to him as *mard-i kamil* (the perfect man), in just this sense, unless the usage is dismissed as mere poetic license (see his poems, *Bang-i dara: Nanak*).

In the earliest and non-canonical writings of Christianity, if we turn further toward the so-called Abrahamic tradition, the theory of the name of God had played a central role, but this is not the orthodox received view (cf pseudo-Dionysius the Areopagite). The etymology of the name Jesus, in this other tradition, would be rendered, 'He who is saves.' It is also known that in the thirteenth century AD there were baptismal rites of Gnostic sects, corresponding to Indian non-dualism, where the neophyte 'puts on' the mystical 'name of Jesus' (Eleazar of Worms).

In the Jewish tradition, on the other hand, according to a well-known Talmudic saying (Shabbath 55a), the name 'Truth' is itself the seal of God, at once real and symbolic. Again, in an early thirteenth century AD text one reads that

> a man was created [artificially] on whose forehead stood the letters *YHWH Elohim 'Emeth*, "God is truth." But this newly created man had a knife in his hand, with which he erased the *aleph* from *'emeth*; there remained *meth* . . . [so] now blasphemy was implied in the inscription, "God is dead" (Scholem 1965: 179f).

In this Jewish tradition, a rite of initiation in the strictest sense is that simply concerned with the transmission of the name of God from master to pupil. For the Kabbalist Jew the great name of God in his creative unfolding is Adam; and the role of Adam Kadmon, primal man, corresponds to that of the perfect man, man as microcosm. God who can be apprehended by man is thus himself the first man.

It was Ibn Arabi (1165–1240) in the Islamic tradition, however, who had influence in India, specially after 1450, when new Sufi centres were set up at Saharanpur, Panipat and Multan. His greatest exponent in India was Shaikh Aman of Panipat, situated near Delhi (d. 1551). Ibn Arabi, the Spaniard, and Jalaluddin Rumi, the Persian, made the 'conspiration' of the sensible and the spiritual the cornerstone of their Islam (Corbin 1970: 322).

Ibn Arabi explained best how it is that 'the spirituals are the kings of the earth'. In his view, prophets and saints and other spirituals are regarded as individuations or particular examples of the 'perfect man', the microcosm, while the universal category or the species 'perfect man' is the complete theophany, the macrocosm, the totality of the divine names and attributes through which the divine essence or the godhead reveals itself to itself, manifested in multiple names and forms, vassals of love (*'abd*). 'Each being has as his God only his particular Lord [*rabb*], he cannot possibly have the whole' (*Fusus,* I). 'The end or goal of love is the unification (*ittihad*) which consists in the beloved's self (*dhat*) becoming the lover's self and vice versa' (*Fusus,* II). Neither the one that is two nor the two that are one, can be lost; it is non-dualism in the sense of unity in the duality and dialogue of the lord of the name and the servant of the name.

The most original contribution of Ibn Arabi is perhaps his non-dualist sense of the question of the true agent of the religious act, the subject of worship, who is none other than God in one or the other of his aspects and names. He says that the created universe is the theophany (*tajalli*) of his names and attributes, which would not exist if the creature, the subject, did not exist. The divine nature (*lahut*) and the created nature (*nasut*), the esoteric aspect (*batin*) and the exoteric aspect (*zahir*), find an exemplary conjunction in the Prophet's person, in which the name of God becomes visible. To know oneself is to know one's lord, because it is this lord who knows himself in you; the name of God is the form under which God reveals himself to himself in that man; this is the non-dualist revelation that we apprehend, and we must meditate upon it in order to know who we are (Corbin 1970: 62, 120, 129, 182f, 295, 300–33).

It is love and suffering that lead us to success in every instance. As long as Maryam, the mother of Jesus, did not feel the pangs of childbirth, she did not go beneath the palm tree (Quran 19: 16ff). This human body is like Maryam, and each one of us has a Christ within him; if the suffering of love willingly rises in us, our Christ will be born (Rumi *in* Corbin 1970: 347).

If I have said nothing directly about the Sikh theory of the name, except to iterate its importance, that is for want of a reliable guide. R.C. Pandeya indeed says that the concept of the name forms the central thesis of the Sikh philosophy of religion, and has studied it from this point of view. It seems that revelation is realized and reciprocally worship is offered equally through *Gurubani*, scripture as the word of the guru. *Sat-nam* or the true name refers to that which is ultimately and absolutely real and true. The *Japji* says that both God and his name are true; the master is real, and the name, word or *logos* uttered by the guru is true. The extent of God's creation is also the extent of his name, which in the sense of *sabad* is the cause of the human world, the cosmos and the individual body. Conversely, the name is also called the God of the Sikh worship, or as the first guru says, 'I am a sacrifice to the name'.

An analysis of the religion of the Sikhs yields a philosophy of language which combines in itself, on the one hand, the Vedic tradition of *sruti*, and on the other hand represents the philosophy of mystic sound developed in the non-Vedic traditions, like Saiva-Agama, Buddhist Tantra and Sahajayana (Pandeya 1975: 82).

4

Sikhism and Islam: History and Society

> That every servant should worship God in the form of his own faith is the law of God's theophany . . . but that he should deny God in the forms of other faiths, upon which he casts the anathema, the *takfir*, that is the Veil.
>
> Corbin 1970: 303

> For the functioning of culture . . . this fact is of fundamental significance: that a single isolated semiotic system, however perfectly it may be organized, cannot constitute a culture—for this we need as a minimal mechanism a *pair* of correlated semiotic systems.
>
> Lotman *et al.* 1975: § 6

The Relational Oppositions

I think that the chief problem of Sikh history is the problem of making an Indian modernity out of medievalism, a painful process of reconstruction, without denying its national heritage. Western modernist writers, liberal and Marxist, persistently misunderstand the problem as being either (*a*) that of mixing up religion and politics, so ultimately causing an explosion detrimental to both, or (*b*) that of setting up a third tradition beside the two existing ones, Hindu and Muslim, in the Indian context. To the contrary, we have explained that it was the specific project or ambition of Sikhism to bring the three spheres of religion, the state and society together, face to face, and so to emphasize the unity of man's estate in modernity, the new configuration of civil society, whereas in medieval India the three were situated back to back, walled off from each other, either for safety's sake or as a matter of principle. The outcome of the project is misunderstood either way, (*a*) or (*b*), as the formation of a state within a state or of a theocracy within a theocracy, and so bound to create conflict. One should view it, therefore, rather as (*c*) a movement towards a new, modern and Indian system of unity in variety, pluralism and civil society, since we have still to learn how to habitually combine, rather than only to separate, a multi-religious nation, a modern pluralist society and a federal secular state.

The traditionalist Muslim writers simply ignore the problem of non-Muslim movements, as is the wont of the ulema, when such are not perceived as a threat to universal Muslim society (*ummat*), or to a Muslim state. The traditionalist Hindu writers, who are called 'communal' in India, declare Sikhism and the Khalsa to have been contingently instituted for the defence of Hinduism, its sword arm in modern history, specially after the so-called militant transformation by the tenth guru at Anandpur, AD 1699, but they do not consider any matters of principle to be involved in the defence, if such it was. The result is that they never deign to speak of Sikhism as being the shield of the Vedas or the sword of the Upanishads, so to say, the sword of gnosis, freedom of conscience and moral discrimination.

As against this line of interpretation, one may cite an historical parallel where there is no Hindu-Muslim question to complicate the issue. Whether its mission succeeded or failed as a whole, the community of the Sikhs, which even today numbers scarcely more than about one per cent of the total population of south Asia, proved itself to be as indigestible to the system of the Mughal empire as did the movement of the Wahhabis of the Najd at about the same time to the system of the Turkish empire in west Asia, at the other end of Islam. The founder of the latter movement was Ibn Abdul Wahhab (b. 1691), a contemporary of Shah Waliullah of Delhi since the two were students together in the Hijaz. The Wahhabis espoused an independent Puritan unitarianism in religion, and in alliance with Ibn Sa'ud, 1744, challenged imperialism and its establishment in the political, military and civil affairs of Arabia; the Wahhabis had conquered Mecca by 1803. Two similar sets of external as well as internal factors were no doubt responsible for this similar outcome in the two cases, Sikhism and Wahhabism, but I shall leave further comparison to some other occasion and to more capable hands.

The problem of religion, politics and history or of their interrelation is always in the background of the mind of all writers on Sikhism. For example, to take an author already quoted, we read of the travels of Guru Tegh Bahadur, the ninth guru.

> The apparent contradiction involved in the reverential attitude of pious Muhammadans, and the skirmishes [of the ninth guru] with Muhammadan soldiery, finds its explanation in the supposition that [*a*] the religious aspect of Sikhism was not antagonistic to Muhammadan ideas, while [*b*] its political aspect provoked the violence of the Court of Delhi. In the present day much the same state of things is recognizable with respect to the Wahhabis. The English Government would never dream of interfering with the religious opinions of that, or any other, sect; but when their doctrines find expression in the subversion of civil authority, the leaders soon find themselves in the Andaman islands (Pincott 1895: 593).

When I say that Sikhism did not attempt to deny its medieval heritage in India, but attempted to change its configuration anew, of course I mean to refer to both Hindu and Muslim. Thus our theological outline so far, in the last chapter, although it has been avowedly concentrated on the relation of Sikh and Muslim, is in general consistent with the simple view that the new way of unitarianism, self-sacrifice and discrimination had no quarrel with the eternal verities of Hinduism as *sanatanadharma*, unless one includes among them the status of Sanskrit as a revealed language and the position of the Brahmin, etc., but only with the system termed *varnashramdharma* and the medieval Brahmanism of the Smritis and the Puranas. This assumes or presumes that the two aspects of religion, the timeless and the historical, can be separated to advantage for a certain purpose, which might be painful to some for the reason that it would demolish the concept of tradition as a single whole.

In its special historical relation to the Bhakti movement and the Sant tradition, which is not disputed by anyone, Niharranjan Ray says that perhaps Sikhism alone was successful in undertaking its movement of social protest and critical reform (1970: 25, 55). It had forged in religion and civil society a middle way between two alternative tendencies. (*a*) The apparent negativism, esotericism and austerities of the non-Sanskritic cults and sects, the religious orders of the Hindu Nathapanthis and Aghorapanthis, the Sahajayani and Tantrik Buddhists and the Jain *sannyasis.* (*b*) What might be termed, in relation to the duly or unduly established system of authority, spiritual and temporal, the policy of 'surrender and survive', associated with at any rate the later phases of the spiritual but asocial, and certainly apolitical, message of Chaitanya, Tulsidas and Mirabai. 'Their aim seems to have been the individual, not the society in any significant sense' (Niharranjan Ray 1970: 61). Indeed, after the mid-sixteenth century, the parallel cults of Vishnu, Krishna and Radha, on the one hand, and of Shiva and Shakti, on the other hand, not to mention the third esoteric cult of Durga or the Devi, equally helped to define the priestly revival in society of the new Smriti-Puranic Hindu reaction to the influence of Islam in south Asia.

Similarly, I would say that the system of Sikhism had no quarrel with the universal and timeless 'spirit of Islam' (*din* or *iman*), which it welcomed, responded to or made its own. But it simultaneously resisted to its capacity wherever it occurred what might be called, following the Sikh writers, the 'way of the Turks', or the spirit of Chingiz Khan, i.e. the medieval worst of Mughal imperialism in politics, and the system of Muslim bigotry, fanaticism and superiority in society, rightly or wrongly claimed to be based on the schools of *shari'at* law (*mazhab*).

I think that a philosophy of language or structural semantic analysis of the usages of the Sikh scriptures and histories would show a system of oppositions, a code of three pairs of related terms, underlying the framework of the message. (*a*) Hindu versus Muslim, applicable to the sphere of religion, which is effectively neutralized by all the gurus by shifting the emphasis from ritual and law to metaphysics, morals and ethics, and so from the two traditions to the one tradition and the tradition of the one. (*b*) Hindu versus Turk in relation to political identity, race or ethnicity, which is differently handled at different places and in different periods, at its sharpest culminating in the martyrdom of the ninth guru, Guru Tegh Bahadur, in defence of freedom of conscience as a universal human right. (*c*) Hindustan versus Khurasan (or Turan), south Asia versus central or west Asia, in relation to locale, habitat or territory, which is polarized as the frontier in history and society, the locus of a conflict of loyalty or of an alternative, a choice of patriotism on the one side or the other.

Historians who find plenty of evidence that 'an undying feud of the deadliest kind' had developed between the Sikhs and the Muslims mean to specially oppose the effects of imperial Mughal persecution, backed by illiberal Muslim opinion, and the warlike spirit of honour of the Jats, many of whom had newly become Sikhs, the national peasantry of the Punjab. This religious and political conflict, we are told by British writers and their followers, led to the emergence of Guru Gobind Singh's political and military Khalsa, defensive and offensive, out of the original pietist and quietist Sikhs of Guru Nanak, himself a Khatri by birth.

Unfortunately, this kind of explanation all too often forgets the parallel evidence that the so-called feudal hill rajas, Hindus of the Himalayas, 'small feudal citadels of an ossified religion' (Niharranjan Ray 1970: 7), with the single possible exception of Mandi, remained from the first to the last implacable enemies of Guru Gobind Singh and the Sikh cause, as implacable as the Mughal imperial authority. The Hindus always seem to have implicitly charged the guru with trying to set up a third tradition of his own, instead of restoring their true one, and so of undermining the society of privilege; while the Muslims chiefly accused him of sedition and trying to undermine the state or the empire—and both no doubt of explicitly or implicitly blaspheming their one true religion.

Martyrdom versus Kingdom: the Fifth Guru

I think that perhaps the world's first martyr of truth and nonviolence was a Greek, Antigone, a European and a woman, best known to us as depicted by Sophocles, *c*. 500 BC. Antigone, who preceded both Socrates and Jesus, wanted the integration of religion and society to be upheld by her freedom of conscience and immemorial usage, the custom of civil society, while Creon the king wished his reasons of state to be separate from, and to override, both religion and society. I will not attempt to decide which of the two points of view is modern for Europe, but it is Antigone's that is closest to Sikhism and the Indian modernity. She had established the truth that no power on earth can really make the self do anything against its nature, except indirectly confer martyrdom on it, which is also the basis of Gandhism in politics.

This abhorred form of the marriage of religion and politics, as it appears to Western modernists, not to put too fine a point on it, is perfectly achieved through the method of *satyagraha* in Gandhism, which is the other Indian modernity in, of and for the twentieth century, and through martyrdom in Sikhism as a creed since the fifth guru. The two are bent upon the model of truth as non-dualism of self, the world and the other in history, religion and society. This affinity made itself visible during the national freedom movement when the two, Sikhism and Gandhism, came together in the 1920s, especially in the Akali party and the *gurudwara* agitation, which was completely non-violent and at the successful conclusion of which Gandhiji sent his telegram to Amritsar saying, 'First decisive battle for India's freedom won. Congratulations' (Ganda Singh 1975: 427). It is no coincidence, I think, that the other most prolonged national movement of non-violence in India should have occurred, equally unexpectedly for some, among the Pathans or the Afghans, under the leadership of Khan Abdul Ghaffar Khan, known as the frontier Gandhi. The British had for long mistakenly identified the Sikhs and the Afghans as splendid examples of the most loyal and/or martial races, so Gandhi's transformation of them would scarcely appear credible, being reduced to a mere seditious plot, and no exalted experiment of truth, but they should have read differently the record of the Mughal empire, and its clash with the discipline and self-sacrifice of their two patriotisms that were one at heart.

Probably the first modern martyr of India—i.e. rather than hero—was Guru Arjun (d. 1606), fifth guru of the Sikhs, who in his mature years steadily and calmly provoked the arrogance of man and the state to reform itself or to kill him, thus establishing an unending line of men and women martyrs for the faith, who are twice daily recalled by all Sikhs in the liturgy (*ardas*), morning and evening. By the example of his life, work and non-violent self-sacrifice or martyrdom, the fifth guru folded up, as it were, the structure of the medieval regime and its intersecting dualisms of status and power, the collective and the individual, exoteric and esoteric, and found for good and all the true centre of freedom, self-rule and self-reform (*swaraj*).

To recover the Sikh and the Muslim relation in history, therefore, we return to begin again with Guru Nanak, who knew very well the ethnic and the political Turk, the Mughal and the Pathan at the turn of the fifteenth/sixteenth century AD. He spoke of the history of his times uniquely among his contemporary spirituals (transl. Macauliffe 1909: I, 117f, 121, 170).

> The Kal age is a knife, and kings are butchers;
> Justice hath taken wings and fled.
> In this completely dark night of falsehood,
> The moon of truth is never seen to rise.

He directly addresses and admonishes the first Mughal emperor, Babur, who had crossed from central Asia over the Hindu Kush passes and occupied Kabul in 1504, and about a generation later defeated the Pathan dynasty of the Lodis at the first battle of Panipat in 1526.

> Deliver just judgements, reverence holy men, forswear
> wine and gambling. . . .
> Be merciful to the vanquished, and worship God
> in spirit and in truth.

Of course Guru Nanak went on further, beyond the history of his times, to address the higher powers.

> The Primal Being is now called Allah;
> the turn of the Shaikhs hath come.
> . . . Babar ruled over Khurasan and
> hath terrified Hindustan.
> The Creator taketh no blame to Himself;
> it was Death disguised as a
> Mughal who made war on us.

> When there was such slaughter and lamentation,
> didst not Thou, O God, feel pain?
> Creator, Thou belongest to all.

For himself and the Sikh community of disciples who were to come to him 'with your head upon the palm of your hand', the religion of Guru Nanak, and not only the religion of Guru Gobind Singh later, is almost as much a way of death as it is a way of life. Guru Nanak frequently speaks of himself in terms of self-sacrifice at the altar of God and his name, and extols the life of 'he who while he liveth is dead'. 'Nanak, in the midst of life be in death; practise such religion.' 'Put the fear of God into thy heart; then the fear of Death shall depart in fear.' 'He is a Qazi who turneth away men from the world, and who by the Guru's favour while alive is dead' (Macauliffe 1909: I, 55, 61, 78, 159, 217, 338).

Now let us try and hear the Mughal voice too in this dialogue in its original accent. Its chief feature, from a Sikh point of view, is that it united the Mughal and the Muslim in domination over the other rather than in fear of God and the love of one's neighbour. It is the voice of Jahangir (1569–1627), fourth emperor in the house of Babur, writing in his memoirs of Arjun, fifth guru in the house of Nanak.

> In Gobindwal, which is on the river Biyah [Beas], there was a Hindu named Arjun, in the garments of sainthood and sanctity, so much so that he had captured many of the simple-hearted of the Hindus, and even of the ignorant and foolish followers of Islam, by his ways and manners, and they had loudly sounded the drum of his holiness. They called him *Guru*, and from all sides stupid people crowded to worship and manifest complete faith in him. For three or four generations [of spiritual successors] they had kept this shop warm. Many times it occurred to me [*a*] to put a stop to this vain affair or [*b*] to bring him into the assembly of the people of Islam.

As it happened, of course, the result was in fact neither of these outcomes, but that Emperor Jahangir tortured to death Guru Arjun at Lahore, capital of the Punjab, partly for his religion and its new scripture, which imperialism found intolerable, and partly for his politics, i.e. because he had once extended brief hospitality and a friendly reception to Prince

Khusrau, the emperor's eldest son, who had then allegedly turned a rebel.

> At last when Khusrau passed along this road this insignificant fellow proposed to wait upon him. Khusrau happened to halt at the place where he was, and he came out and did homage to him. He behaved to Khusrau in certain special ways, and made on his forehead a finger-mark in saffron, which the Indians (Hinduwan) call *qashqa*, and is considered propitious (*tika*).
>
> When this came to my ears and I clearly understood his folly, I ordered them to produce him and handed over his houses, dwelling-places and children to Murtaza Khan, and having confiscated his property commanded that he should be put to death (Jahangir 1909: 72f).

It is sociologically important to remember that this was done in spite of the historically proven attempt at intercession or mediation, but which proved unsuccessful, by the saintly Miyan Mir of Lahore, a Muslim faqir (d. 1635), who was held in the highest regard by both sides in this relation of martyrdom *versus* kingdom.

An Indian historian, Indubhusan Banerjee, who had acquired first hand knowledge of the Persian as well as the Punjabi original sources, concludes that Guru Arjun's participation in the prince's rebellion was neither very deep nor very extensive. The underlying causes of provocation were rather different; he enumerates three of them and they are more or less true and not to be completely denied. Indeed, it was Guru Nanak's view also from the beginning that God is the only true king, and all earthly kings are more or less false and usurpers.

> . . . The charges against the Guru also included the allegations [*a*] that he called himself the "true king" (*saccha padshah*) [i.e. spiritual and temporal, "the true king of the world", as later in the Sikh chronicles of the sixth guru, transl. Macauliffe 1909: IV, 18], [*b*] that he had established a large organization with the intention of making war upon the Emperor, and [*c*] that he had compiled a book which blasphemed both the Hindus and the Musalmans (Indubhusan Banerjee 1962: II, 2, 20).

The life, work and death of Guru Arjun perfectly represent all that Guru Nanak had founded and anticipated, i.e. the three elements, attitudes and relations of the Sikh and the Muslim dialogue in modern Indian history. (*a*) The love of the Sufi (*tariqat* and *haqiqat*) in mutual attraction; (*b*) the hatred of 'the Turk' (*hukumat*) in mutual repulsion; and as further analysis would show, (*c*) the impartiality, neutrality or indifference of the purely legalist ulema (*shari'at*). These are the attitudes that continuously shaped the discourse of freedom and sovereignty in the house of Babur as well as the house of Nanak, but I think that their rival claims in relation to history were respectively those of old medievalism and new modernity in India. The same structure of elements, relations and attitudes had also simultaneously determined the internal reorganization of Sikhism at the time.

For, on the one hand, Guru Arjun had personally edited and compiled the Adi Granth, the holy scripture of the Sikhs, including in it the verses of Kabir (b. 1398 or 1440), Farid I or Farid II and of Bhikan the Sufi (d. 1574) as sacred along with the word of his predecessors. At his invitation, the same Miyan Mir, faqir of Lahore, had probably laid the foundation stone of Harimandir Sahib, the golden temple at Amritsar, or had been at any rate present on the occasion; the holy book was formally installed there in 1604. The land of the temple was an original gift of Emperor Akbar in 1576, made during what historians describe as his liberal interregnum (see Gurcharan Singh 1976, which also includes a portrait of Miyan Mir).

Mir Muhammad, popularly known as Miyan Mir, was later appointed by authority as tutor to the liberal and scholarly but ill-starred Mughal prince, Dara Shikoh, 1615–59. This might explain the latter's known friendly relation with Guru Hargobind, 1595–1644, the next guru of the Sikhs. Miyan Mir was a shaikh of the Qadiri order of faqirs, perhaps the oldest ascetic and mystical order founded in the twelfth century by Sayyid Abdul Qadir Jilani of Baghdad (1078–1166), surnamed Pir-i Dastagir, 'the patron saint', under which title the poor and the working class put themselves under his protection in all parts of the Muslim world. The Qadiri slowly and gradually became the most popular religious order among the Sunnis in many places of Asia and Africa. After *c.* 1450 and by 1500 many of the Sunni *maulawis* of the north-west frontier of India were members of this Sufi order. Its members were said to sit together for hours at a time uttering the words, 'Thou art the guide, Thou art the truth, there is none but Thee!' This is an example of the popular Muslim invocation of the names of God, *da'wat-i asma*. Its shrines in the Punjab, beginning with the foundations at Multan, Uch and Lahore in the time of Humayun or earlier by Sayyid Muhammad, known as Bandagi Muhammad Ghaus (d. 1517), were rivalled or exceeded only by the Chishti order.

At the time that Sikhism took its birth, it seems that in general the Sufis of Hindustan, Sunni as well as Shiah, befriended non-Muslims or 'unbelievers', according to the Sufi principle of *talif-i qulub*, the stringing together or joining of hearts, which came very close to the heterodox or esoteric 'equality of all religions'. The orthodox ulema, on the other hand, properly limited themselves to the other parallel task of making good Muslims out of nominal Muslims, chiefly through regulating the law of exoteric ritual and society, the practical minimum Muslim code of conduct. The ulema remained largely indifferent to non-Muslims, the relation with whom they advisedly left to be regulated by the state as the third party, whether Mughal or later British, and even nowadays.

On the other hand, one must admit, however, that Guru Arjun himself laid aside the garments of a faqir; indeed, he wore a sword in his belt, and he turned the voluntary offerings of his Sikhs into a treasury of the community (as still today), which enabled him, among other things, to take soldiers as well as officials into the Sikh employ in the future. The fifth guru may not have meant to give offence to, much less to wage war upon, the emperor, but he was effectively urging the claims of pluralism versus imperialism or one single central rule in culture as well as power, religion, civil society and political economy, as a matter of conscience.

His infant son and successor, Guru Hargobind, the sixth guru of the Sikhs, put the challenge of pluralism, freedom and responsibility in matters of religion, civil society and politics even more plainly. It is said that he refused upon his assumption of the guruship after his father's martyrdom the offer of the *seli*, the headgear and necklace of renunciation as worn by Baba Nanak. He adopted novel usages and wore now his turban with a royal aigrette and carried two swords, one on his right and the other on his left, signifying as 'impersonal, indivisible and continuous' the unity of the spiritual and the temporal powers, the Piri and the Miri of the Islamic tradition, and the *bhakti* and the *shakti* of the Hindu tradition. In AD 1606 he also caused to be built and took his seat in the Akal Takht, presence of the timeless in the temporal, situated across the causeway from the Harimandir, specially devoted to the service of the word, conjoining the two parts of the golden temple face to face in the new tradition of the city of Amritsar, the pool of nectar, immortality or salvation. It is the usage and custom of the golden temple that the single volume of the holy scripture, the Granth, is to be daily installed, perused and addressed during the day in the Harimandir, and made to lie at rest during the night in the Akal Takht, the two moments of transition being attended by prescribed 'rites of passage', morning and evening.

Such were the second fruits then, to vary a phrase, of the project of Sikhism exactly a hundred years after the first guru instituted it in the Punjab, based upon the philosophy of the name of God and a perfect discipline of self-abnegation and self-sacrifice. The fifth guru matured the project and added the conflict of martyrdom versus kingdom in defence of pluralism to the Indian modernity at the turn of the sixteenth/seventeenth century AD. This is a constant underlying theme ever since then with the community and perhaps nothing in his life did so much for the Indian nation and modernity, from a Sikh point of view, as did Mahatma Gandhi's own martyrdom, which he apparently sought, at the hands of an outraged orthodoxy in the twentieth century.

After Anandpur, AD 1699

From a sociological point of view, the mutual struggle of classes, strata and groups remains somewhat obscured to us at this juncture, when the Jats freely entered the Sikh fold. The very general fact is that Sikhism always led the productive sections of society in the Punjab, defined as the land of the five rivers of the Indus valley lying between Delhi and the north-west frontier, in the occupations of agriculture, commerce and the crafts, especially the artisanate, rural and urban. Whether it was from the point of view of the Jat or the Khatri, call them tribes or castes, not to mention the Harijan, the Untouchable or the Dalit, sweepers and leather-workers, the social class background of Sikhism and the Khalsa was eventually a settled combination of the national peasantry, the artisanate and a sort of free knighthood of chivalry, i.e. independent rather than following either the landed gentry or the rajas. This is exactly the same class combination that failed to achieve its reformist objective in early modern Europe at about the same time first in the great peasants' war in Germany, 1525, and equally in the second wave of the radical Reformation, signified by the fall of the non-violent Anabaptists of Münster, 1535.

Thus in Sikhism material relating to the mercantile life and the agricultural life, Khatri and Jat, is frequently employed to explain the message of the Adi Granth. For example, the same *Japji* of Guru Nanak ends with the following penultimate verse that relates the artisan's work to the spiritual work of the Sikh (Macauliffe 1909: I, 159, 217).

> Make continence thy furnace, resignation thy goldsmith,
> Understanding thine anvil, divine knowledge thy tools,
> The fear of God thy bellows, austerities thy fire,
> Divine love thy crucible, and smelt God's name therein,
> In such a true mint the Word shall be coined.
> This is the practice of those on whom God looketh
> with an eye of favour.
> Nanak, the Kind One by a glance maketh them happy.

In the same tradition, Guru Gobind Singh has finally added in his writings a third dimension of terms relating to the life of the fighter, warfare and weapons, even as the names of God. He expresses the exoteric and the esoteric aspects conjointly so that the sword of the Khalsa remains that of the spirit and gnosis, knowledge and discrimination, not merely of the soldier's avocation. Whether it is single-edged or double-edged (the *kirpan* or the *khanda*), it is a cutting and not a piercing weapon, a weapon to part the veil of the world (*maya*) and so to look beyond it. Similarly, the Sikh custom of slaughtering animals for food by killing with one stroke, which the tenth guru probably introduced or sanctioned, is so that the creature may die without experiencing fear (*jhatka*). This is as distinct from the Muslim custom of eating only *halal* meat that is obtained by the complete and therefore slow separation of the flesh and the blood of the animal, while reciting the great name of God or the name of the great God, *Allahu-akbar*, over it as a sacrifice, in the Semitic tradition.

> He who taketh the sword of knowledge and
> wrestles with his heart;
> Who knoweth the secrets of the ten organs of action and
> perception and of the five evil passions;
> Who can knot divine knowledge to his mind;
> Who maketh pilgrimage on each of the three hundred
> and sixty days of the year;
> Who washeth the filth of pride from his heart—
> Nanak saith, he is a hermit [as well as husband].

The torture and execution or the voluntary martyrdom of Guru Arjun, fifth guru of the Sikhs, took place at Lahore in the year 1606; and the whole pattern exemplifying the grammar of conflict was manifest in the similar imperial execution or the voluntary martyrdom at Delhi of Guru Tegh Bahadur, ninth guru of the Sikhs, in 1675. One may add for a third time in a hundred years, the execution and the martyrdom for refusing to deny their faith, of the two infant sons of the last guru, Guru Gobind Singh, aged seven and nine years, by the viceroy of Sirhind in 1705. All these tragic events were associated, an historian concludes from the one side, with the 'meanest and cruellest barbarities of the medieval world' (Niharranjan Ray 1970: 38); but what did they signify of Sikhism and its modernity?

We have already briefly discussed how on the new year's day of Baisakhi at Anandpur, 1699, Guru Gobind Singh instituted the Khalsa as a society for salvation through the baptism of the spirit and the sword of gnosis, and conferred on it the freedom and the responsibility of both spiritual and temporal self-rule (chapter 3). 'I shall ever be among five Sikhs. Wherever there are five Sikhs of mine assembled they shall be priests of all priests. Wherever there is a sinner, five Sikhs can give him baptism and absolution' (Macauliffe 1909: V, 189). This completed the edifice of Sikhism as the work of the gurus; and the mutual identification of Guru, Granth and Panth through their reciprocal embodiment and participation. After that time, the project must be continued as the work of the Sikhs, the own responsibility of the Khalsa, of course, under the guidance of the scripture and the gurus' ever present example.

Altogether, this final event at Anandpur led the tenth guru to ask every Sikh, including his four sons, to belong to God, to try and reduce the inner and outer ills of existence, individual and collective, and to forfeit life rather than faith. Such were the third fruits of the project of Sikhism at the turn of the seventeenth/eighteenth century AD. The tenth guru's completed theology was thus addressed in prayer to God as the threefold beloved lord, 'Protector of the saints, Friend of the poor, Destroyer of tyrants.' The free Sikh of the new commonwealth was ever to try, not like the soldier and the hero to be stronger than the other, but like a martyr or a saint to be stronger than oneself.

The way of unitarianism, self-sacrifice and the society for salvation was to ever remain the way of martyrdom versus kingdom, and from this point of view the so-called Sikh kingdom of the Punjab, which flourished under Maharaja Ranjit Singh, was aberrant and not the acme of Punjabi nationalism as some people would have it. Indeed, he allowed to fall into desuetude after 1805 the practice and the sovereignty of the *gurumatta*, a resolution of the general assembly of the community's equal segments (*misals*), held by custom at the Akal Takht of Amritsar. Anyhow, a few years after the event at Anandpur, we happen to have independent foreign testimony of the spirit of Sikh self-sacrifice made in the lord's cause, as against the militarism of which they stand unjustly accused.

The British mission that came from Calcutta to the capital of Delhi in 1715 to petition the Mughal emperor for some privileges has left us a record of how it saw a procession of some 800 live Sikh prisoners, apart from 2,000 fresh bleeding heads of the rebels borne aloft on poles, marched through the streets of Delhi to the place of execution. It concludes with the observation that 'the Sikhs vied with one another for precedence in death' (Macauliffe 1909: V, 169ff, 252, 328).

> While the executions were in progress, the mother of one of the prisoners, a young man just arrived at manhood, having obtained some influential support, pleaded the cause of her son with great feeling and earnestness before the Emperor. . . . Farrukh-Siyar pitied the woman,

and mercifully sent an officer with orders to release the youth. She arrived with the order of release just as the executioner was standing with his bloody sword upheld over the young man's head. When she showed the imperial order the youth broke out into complaints, saying, "My mother speaketh falsely: I with heart and soul join my fellow-believers in devotion to the Guru: send me quickly after my companions." Needless to say his request was cheerfully granted.

When all of this is said and done, and imperialist reasons of state heard and considered, Abdul Majid Khan is surely perfectly right to insist that, even at the height of the conflict, Guru Gobind Singh himself had found 'admirers and devotees', not to say disciples, among the Muslims until the very last—and not only among the Sufis, who might have left the world, but also among all classes of the nobility and the common people of the Punjab. We set aside here awhile the hundreds of Afghan soldiers in the guru's regular service, in case they might be dismissed as merely mercenaries or opportunists.

He is thinking of Sayyid Bhikan Shah of Kuhram, of the two *nawabs* Rahim Baksh and Karim Baksh, Pir Arif Din of Lakhnaur and Pir Budhu Shah of Sadhaura, Ghiyasuddin the faqir, and of Ghani Khan and Nabi Khan, two very ordinary people known to Sikh history as the ones who saved the life of the tenth guru in the wilds of Machhiwara, passing him off under questioning to escape the imperial troops as the Pir of Uch, 'the high master', by a sort of a linguistic ambivalence.

Most of all, however, the Sikhs honour the memory of Sher Muhammad Khan, *nawab* of Malerkotla, who had protested against the order of execution of the two children in 1705, and written to the emperor that this order seemed to him contrary to the law of Islam and the Prophet. Several independent foreign witnesses have testified that, even now during the mutual hostilities of the partition of India (1947), Malerkotla and its environs formed a protected sanctuary for hapless Muslims in the eyes of even the most fanatical Sikh (see Abdul Majid Khan 1967b: 89ff; and C.H. Loehlin 1967: 112ff).

It might be objected in general by a Hindu as much as a Muslim that, leaving aside the exceptions that were few and far between, the Sikhs were essentially the sword arm of the Hindus versus the Muslims, specially from the institution of the Khalsa up to or down to the partition of India, if not until today. In my reading of the structure of theology as well as history, synchrony and diachrony, and even if my hypothesis of the transformation from medievalism to modernity were to be rejected, the answer is somewhat like this. The Sikhs have some similarity of religion and culture to the Muslims, which we have honestly delineated, and for that very reason the two would be sometimes in the relation of opposition, competition or conflict, in effect even to the nadir of being willing partners in a pact of the suicide of values, so degenerating into the bloody clash of rival fanaticisms, but rule over others is nowhere on the agenda of Sikhism. On the other side, the very difference of religion and culture, wide or narrow, of the Sikhs from the Hindus could join the two together in a relation of complementarity in history, as for instance in the martyrdom of Guru Tegh Bahadur trying to protect the custom of wearing denominational marks by the latter as a right of conscience. Sharing the 'unity of origin' with the Hindus, so to say, and the 'unity of goal' with the Muslims, or with some of them, Sikhism could produce for that very reason a negotiation, mediation or reconciliation in history, and without introducing any false syncreticism or homogeneity.

An Historian's View, *c.* 1500–1947

At the very least, therefore, Sikhism as method and praxis attempted to integrate the medieval dualism of the pursuit of gnosis, Bhakti, Sufi or Sant, and what might be called the material interest of the world, specially the labour of production, property and the family life of reproduction. The project of Sikhism, as I have argued, was that of a society for salvation, by its free nature and circumstance unitarian in religion, vernacularist in culture and democratic in politics. This society is a sacred and joint construction of the guru and the Sikh, and it is not a pragmatic or a contingent one; it has a history of freedom and not of determinism. It is the conjoint product of reciprocal embodiment and mutual participation of the guru and the Sikh, and it is set up in response to the three original commandments of the founder and first guru: work or labour (*kirt karo*), worship of the name (*nam japo*), commensality or sharing (*vand chhako*). The first institution of Sikhism is the worship of the name of God in and by the local congregation; and all three commandments together constitute it as a society for salvation, aiming to achieve self-realization through self-abnegation and self-sacrifice in this life and world, sealed by the signature of the serene and unshorn martyr.

The sum of Sikh history is simply that its mission of religion and its commandments are to be achieved or realized through society, the guru and the Sikh, and not through the state, even under Maharaja Ranjit Singh (1780–1839), the secular monarch of the Punjab. The real and true symbol of Sikh history, therefore, is the city of Amritsar and the golden temple, the Harimandir Sahib and the Akal Takht, spiritual and temporal, facing one another and joined by the causeway and reflected in the pool of immortality. It is surely not the royal and ancient city of Lahore, although the latter had to be won by the Sikhs in slow degrees from the Afghans finally in 1799, and was not lost to the British until 1849.

One may now recast, in the light of this other evidence, the conclusion of the historians, Indian and foreign, who write that the story of Sikhism ran on chiefly pacifist lines, pietist and quietist, from his assumption of the ministry by Guru Nanak in 1499 or 1507 until the compilation of the Granth was completed by Guru Arjun in 1604. Surely, the system of Sikhism gradually detached itself from the medieval tradition of Hinduism as a system of castes and sects, ritual and society (*varnashramdharma* versus *sanatanadharma*). It developed social ideals and institutions of its own and the Sikh Panth, the new confraternity, developed its form and significance in the Punjab, the land of the five rivers, and also beyond it. Its first century was that of the Sikh and the Hindu relation, therefore, rather than that of the Sikh and the Muslim relation. The system of Sikhism was evidently not in conflict with the religion of Islam (then or at any other time), nor yet with the Mughal state that was then dominant in India.

Yet the structure of the Sikh religion, its unity of the godhead (Guru), the scripture (Granth) and the community (Panth) was already well and truly laid in its philosophy, self-identity and action.

It would thus appear that as, on the one hand, love of mankind was narrowed down to love of the followers of Nanak, on the other, service to the Guru was expanded into service to Sikhs in general, and the Sikh brotherhood was ushered into being. . . . On the one hand, we have the unconditional surrender of the Sikh to the Guru; on the other, the almost equally unconditional deference of the Guru to the will of his disciples (Indubhusan Banerjee 1963: I, 246f).

Thus, 'the guru is the Sikh and the Sikh is the guru' in identity as well as function.

. . . The vesting of the spiritual leadership of the community in the Guru Granth Sahib and of the temporal leadership in the Khalsa itself was the culmination of a process that had long commenced in Sikhism (Indubhusan Banerjee 1962: II, 119).

The political and military Khalsa, if such it was, initiated by the tenth guru (d. 1708), and whether it was predestined to be so or not, emerged triumphant in the Sikh and the Muslim relation during the second century of Sikhism. The new Khalsa of Guru Gobind Singh is best explained, writes Indubhusan Banerjee, as the synthesis of which the earlier Sikh was the thesis and the Jat the antithesis. It admirably combined, therefore, the spirit of God-mindedness with the spirit of rectitude, self-sacrifice and discipline; and it was an ideological corporate body.

He continues and sums up for the third century of Sikhism that, although they were squeezed, persecuted and massacred by the imperialist Mughals from the centre, on the one hand, and constantly pillaged by the tribal Afghans of the frontier, on the other hand, the Sikhs of the Punjab yet succeeded in the struggle to possess a territorial sovereignty of their own in the eighteenth century AD, which freed the land of the five rivers from Delhi up to the northern and the western frontiers. From the time of the first guru up to the present, in other words, the Sikhs have always sided with south Asia as against central Asia or west Asia, with Hindustan as against Khurasan or Turan, as a matter of principle, the patriotism of territoriality or the concept of a homeland. 'It is important to remember that the British conquered the Punjab not from the Afghans but from the Sikhs and, when they took over, they found the frontier intact, with every inch of Indian soil inside it' (Indubhusan Banerjee 1962: II, 155).

The history of the British conquest of the Punjab and its dependencies and the later British period, i.e. the story of 1848 and the century that followed after the institution or the interlude of the Lahore durbar (1799–1848), is well known and need not be retold. Perhaps it will show the same kind of structure and dialectics that we have anticipated, and which led General Sir John Gordon, the Englishman, to conclude that, in the history of the British empire, 'None have fought more stoutly and stubbornly against us, none more loyally and gallantly for us, than the Sikhs' (1904). This British century came to an inglorious end with the terrible tragedy of partition of the Punjab (1947), parallel to that of Bengal at the other end of Hindustan, when the Sikh and the Muslim absolutely separated their territories, with the sole exception of Malerkotla which remained in the new India.

The other half of the story, specially Sikh political participation in the national freedom movement and the Sikh contribution to Indian social and economic developments before and after independence (1947), is not so well known but I cannot enter upon it here.

I have no objection if someone were to argue that all of this story, *c.* 1500 to 1947, relates to the 'superstructure' of religion, the state and civil society, and wishes to produce another periodization or a corresponding dialectics of the 'infrastructure', providing only that the notion of structure is given its full value along with that of dialectics, and synchrony is respected as much as diachrony. Even the best historians who attempt to survey the general field are still too much inclined to see all history in terms of the determinist cause and effect chains of a sequence of events, stages or influences, or worse in terms of the 'interaction' of two or more such separate and different sequences. India has for them no principle of motion, dialectics or contradiction of it own, civil society is reduced to political economy and modern history to our response to others, chiefly British. It is assumed that the history of religion and the history of politics were, or should have been, two separate and different histories, in spite of all cogent reminders of M. Mujeeb and Niharranjan Ray, for example, that the 'organic fusion' of the spiritual and the temporal was sought for existence as well as salvation, thus making reference to only one history, the way of life (and death) of man in society and of man and society in God.

We have attempted, subject to limitations, to show a structure and dynamics of synchrony as well as diachrony in the relations of Sikhism, Hinduism and Islam, using the same terms of discourse in the aspect of commonalty, mutuality and sharing as well as in the aspect of conflict. The religion and history of India, howsoever one may periodize them, are not only to be allocated among two or more separate and different traditions, but also to be regarded as offering the conjoint examples, questions and answers of the interrelation of religion, civil society and politics viewed as a general problem of the human species.

As Gandhi emphatically said, 'Politics concern nations and that which concerns the welfare of nations must be one of the concerns of a man who is religiously inclined, in other words, a seeker after God and Truth.' 'Therefore, in politics also we have to establish the Kingdom of Heaven' (*Young India*, 18 June 1925). There can be no better summary than this of the cause of an Indian modernity, the cause espoused by the Sikhs in history and society in three steps of continuity or discontinuity, beginning with (*a*) the foundation of Sikhism in a plural society as the one tradition and the tradition of the one, in *c.* 1500. (*b*) Aspiring for participation in history and politics through the society for salvation, self-sacrifice and ethical discipline, which we may now back-date as the moment of revolution to 1606, seeking non-violent martyrdom versus the kingdom of powers. (*c*) The embodiment of reciprocity in the relation between the 'master of the name' and the 'pupil of the name', the guru and the Sikh, individual and collective, finally achieved in institutional form through the event of 1699.

Pluralism and Civil Society

One lesson of our study so far, as we have sought to combine synchrony and diachrony, for me is that the plurality and the variety of traditions is a great human achievement in the discourse of history and theology or ideology; and that it has always formed part of our most precious heritage in India, medieval and modern.

It is not susceptible of future recognition and development, however, within a framework of either domination, stratification and hierarchy, or alternatively of a falsely imagined unity through uniformity, homogeneity and centralization. There is indeed a complex problem of similarity and difference, proximity and distance, or of interaction and innovation, competition and co-operation, in the interrelations of Hinduism, Sikhism and Islam, not to mention the others. It should be seen, however, within the philosophy and the institutions defined in the Indian modernity by Mahatma Gandhi under the heading of the 'equality of all religions', i.e. all equally true and truly equal as well as all equally imperfect, and therefore each of them capable of self-reform as well as self-rule (*swaraj*). It is useless to try and distinguish in the study of comparative religion between the so-called revealed religions and the others, whatever they may be named, since God cannot reveal a religion, but only himself, and all religions are in some sense man-made in history and society.

Apart from, or along with, the hardy perennials of domination and exploitation, death and destitution, affliction and alienation, the problem for the world today is perhaps the reconciliation of difference with equality in civil society. This surely cannot be achieved within a framework of the (permanent) majority and the minority, superordination and subordination or the centre versus the periphery, but will require embodiment of and participation in the logic of pluralism, mediation of the one and the many, which we call in India 'unity in variety'. It is hardly necessary to add that to this attitude in religion, for example, would correspond a true federalism in the state, even if under the Mughals and the British, and real tolerance and purity in civil society and the economy, which is the modern Indian way of combining as well as separating a multi-religious nation, a modern pluralist society and a federal secular state.

We have considered above Sikhism both as a product of Indian pluralism and as making a contribution to it in synchrony and diachrony. One should now explain, according to the rules of semiological method, the background of the analogy of language.

Linguistic diversity is an established way of life in India; . . . which is, in fact, one aspect of cultural pluralism. People in a multilingual country develop ways of speaking in which they can maintain their own language and still communicate with speakers of other languages; they become bilingual or multilingual. Languages are not employed parallelly in all spheres of activity, . . . the patterns of use of languages in a [single] bilingual community are largely complementary.

All languages are equal, but different languages have different roles in society, and have different positions in the education system. . . . That each speaker has access to the other's variety (e.g. Hindi or Urdu) and also to the common variety (Hindustani) underscores the integrative aspect of such variations [as against uniformity]. The assertion and acceptance [of identities] are reciprocal and mutual. In this way, different social groups can maintain their separateness on the one hand and express their togetherness on the other. The face-to-face communication allows for more [strategies of] negotiation than any other context.

[Pluralist] integration is thus different from [*a*] "melting pot" on the one hand and [*b*] segregation on the other. Melting pot results into complete assimilation with the dominant group, a merger of identity. Segregation results into isolation and the tensions thereof.

Hindi and Urdu, in the language of structural linguistics, form one set of correlation; it is an opposition in which one can only be explained in the context of the other: Hindi and Urdu are two sides of the same coin: if you have one then you have to accept the other. . . . Urdu identity becomes more relevant [in its diacritical function] only in the context of Hindi rather than in the context of any other language.

Though the institutional affiliation of Urdu is with Persian literary tradition and that of Hindi with the Sanskrit literary tradition, Hindi and Urdu literature have been developed by the same population in the same region—Hindu and Muslim writers have contributed to both these varieties. As in various other spheres of creative arts, music, painting, dance and drama, so in writing, there has been a remarkable reciprocity. . . . Persian scholarship flourished alongside the traditional Sanskrit scholarship for a long time in the Northern courts . . . [and] for the first time in the history of Indian languages, a common *lingua franca* [Hindustani] becomes the language of the court, administration, law and order though spread [and varied] over a vast territory. . . . Linguistically, Urdu and Hindi are recent [alternative] standardisations of a medieval [or early modern] *lingua franca*.

Our linguistic issues are not multiplicity of languages, social variation and varieties, but of the gaps between the colloquial and the literary standards (Pandit 1977: 3f, 37, 47, 50, 56, 60).

Functional multilingualism in India, therefore, makes the common usage prevail over inherited tradition and perhaps this

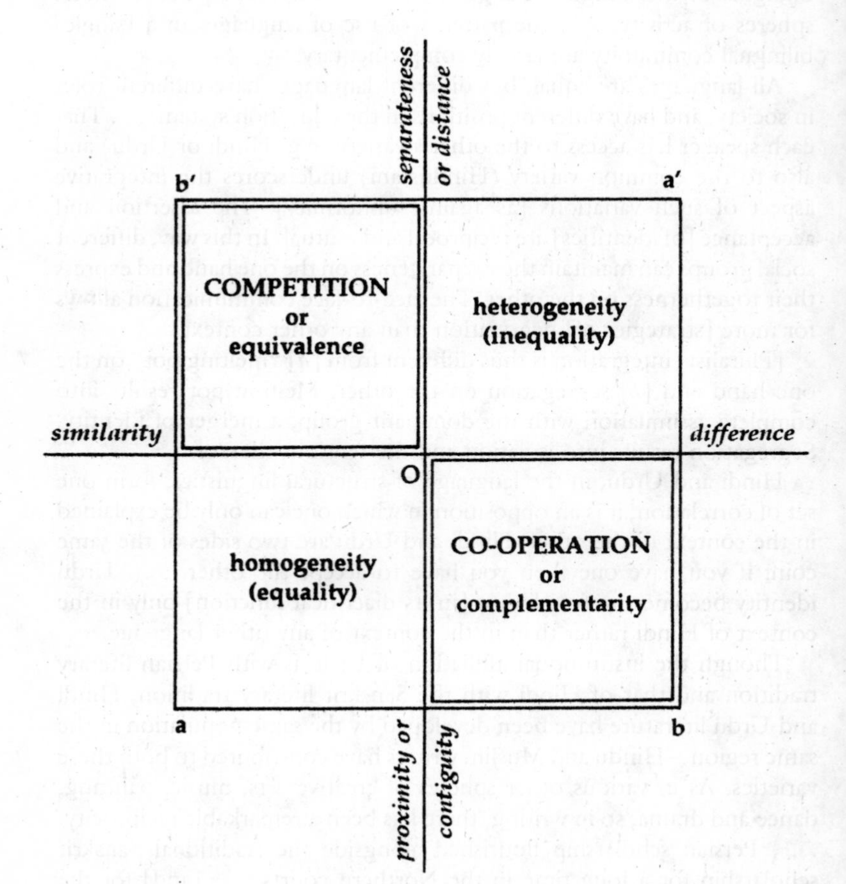

FIG. 4 Structure of National Pluralism

is the normal and the proper condition of modernity. It is inclined to keep its differences, or some of them, contextual and domain-specific in relation to family, neighbourhood, bazar, formal education, work, administration, worship and play, and distributed, so to say, on the horizontal plane as well as on the vertical plane of society and culture. Thus several independent studies affirm that languages, for instance, grow and develop, not by purity, insulation and officially-sponsored standardization, but by ongoing contact, reciprocity and exchange with other languages in a plural setting, whether national or international.

The European writers whom I quote on the subject, e.g. French and Slavonic, have duly recognized but two tendencies at work in history and society, centripetal and centrifugal, and those two opposed and unmediated, whereas I think that India will show us a field of four possibilities to make up the logic of pluralism, the one and the many.

> Humanity is constantly struggling with two contradictory processes. One of these tends to promote unification, while the other aims at maintaining or re-establishing diversification. The position of each era or of each culture in the system (the orientation in which it finds itself engaged) is such that only one of the two processes seems to have a meaning. The latter is seemingly the negation of the former, . . . we are dealing with two different manners of *making oneself* (Levi-Strauss 1977: 361).

> In the union of different levels and subsystems into a single semiotic whole—"culture"—two mutually opposed mechanisms are at work: (1) the tendency toward diversity [heterogeneity]—toward an increase in differently organized semiotic languages, the "polyglotism" of culture; (2) the tendency toward uniformity [homogeneity]—the attempt to interpret itself or other cultures as uniform, rigidly organized languages (Lotman *et al.* 1975: § 9).

All variation of custom or usage, taking the analogy of language, which gives the alternative appearances of distributions that are either on a vertical plane (stratification) or on an horizontal plane (segmentation), can be resolved and united into a single systematic field of national pluralism or 'unity in variety'. The two diagonals on fig. 4 represent respectively the theory of social stratification, equality and inequality (*aa′*), versus the theory of division of labour, as it is called, segmentation or complementarity (*bb′*). We can resolve both of them simultaneously into the two fundamental axes as shown, horizontal and vertical, i.e. respectively similarity/difference and proximity or contiguity/separateness or distance, and term the whole reconstituted single system or unified field of identity, classification and exchange, 'unity in variety'. This is the structure of national cultural pluralism, as explained further elsewhere, and one of its remarkable consequences is the convergence as well as divergence or emergence of varied, not to say disparate, modes of articulation and communication (Uberoi and Uberoi 1976).

P.B. Pandit's new contribution to sociolinguistic theory lies, then, in his studies of what we have termed the horizontal plane or axis of speech varieties. We have inferred that multilingualism, taken as a characteristic Indian and human way of life, is constantly sustained, made functional and integrated by virtue of the simultaneous operation of two general tendencies or principles: (*a*) the distribution of differences into complementary, rather than only into competitive, domains of activity or contexts of situation; and (*b*) the convergence of underlying structures in history and society through the free human act of communication and exchange. The first principle is a kind of self-control on differentiation within a system, while the second is a self-control on variation or diversification between two or more systems. We should say that, under a regime of pluralism, all differences are negotiable in civil society because no reality or truth falsifies another, and that the strategies and the dialectics of negotiation, whether in amity or enmity and harmony or conflict, together explain the processes of the convergence of underlying structures and the distribution of differences into complementary domains, in effect thereby producing neither simple homogeneity nor heterogeneity but a new non-dualist axis of mediation.

In the next chapter, I shall attempt to give three examples from Shiah Islam, Sikhism and Gandhism of the good combination of religion and politics in the theatre of civil society, i.e. when both of them are in the service of self-rule and self-reform. Here we may conclude the evidence and argument so far presented in my own terms, although the Western modernists will say again that, from about AD 1500 up to the mid-twentieth century, the first battle of Panipat up to the partition of India and of the Punjab, the relation of Sikhism and Islam in Hindustan, if it did constitute a single history, simply produced a misguided attempt to set up a theocracy within a theocracy or a state within a state. It is as if one were to narrowly concentrate one's view exclusively on the ways of princes and priests, the sphere of *hukumat* or statecraft and the sphere of the ulema, legists of the *shari'at*, misconceived as theologians, and forget the ways of the common people, the martyrs and the saints, on either side.

I have in fact argued that, in the sphere of truth or *haqiqat*, the two (or more) traditions were able to celebrate the confirmation of one revelation by another as well as the worship of a God who was the maker of a plurality of forms out of himself for mystical reasons of his own, to reveal his virtualities in actuality to himself. The Islamic concept of *tauhid* assumes and implies the 'unity of origin' as well as the 'unity of goal' of the forms, which always seek to return to him along the way of loving self-sacrifice, self-realization and obedience (*tariqat*). This is opposed, for instance, to the worship of a God who is the author of two principles, architect of two mutually estranged spheres, the spiritual and the material, truth and reality, or the soul versus the cosmos, originally created by him out of chaos or nothing. It is justly said that creation, flowing forth from God as the world, returns to God as man; that is the cause and the law of our being; and every sacred society or quest must take it as its first task to embody the process, individual and collective, and participate in it, specially under what is called a secular state.

'Islam's distinctive contribution to India's national culture', as Gandhiji wrote, 'is its unadulterated belief in the oneness of God and a practical application of the truth of the brotherhood of man for those who are nominally within its fold' (*Young India*, 21 March 1929). The same can be said of the contribution of Sikhism to the national culture with equal justice, and this similarity alone should suffice for the justification of our theme, Sikhism and Islam, but I may add three points on the relation of religion and politics, the project of Sikhism and the Indian modernity.

(*a*) I may conclude that politics, in the sense of the self-management of society, far from being considered secular and relegated to a secondary position, status and role in Sikhism and Gandhism, is to be taken religiously seriously. This is to be done directly in the sense of Rama Rajya or Khalsa Raj, i.e. not through the state but by the self-rule of the community under divine guidance and the gurus' example. The institution of the state itself remains secular within its limits, i.e. without any divine right of kings, as it always was in the concept of Islam as well as Hinduism in India. In the Indian modernity, the state must learn to live and let live under a regime of pluralism, and even to tolerate other sovereignties, free and responsible, besides its own in society. It need not and must not defer to a higher religious authority, as was said of the ancient period, nor yet remain partitioned from the sense of the sacred, as under the medieval regime.

GURU	:	GRANTH	:	PANTH
		::		
FIRST GURU	:	FIFTH GURU	:	TENTH GURU
		::		
Theory of the name	:	Theory of martyrdom *v.* kingdom	:	Society for salvation
		::		
Sultanpur/ Kartarpur	:	Lahore/ Amritsar	:	Anandpur/ the Khalsa
		::		
THE FRONTIER	:	THE CENTRE	:	THE FUTURE

(*b*) The first guru of the Sikhs was, as a seeker after and teacher of truth, non-dualism and self-sacrifice, surely a man of the frontier and the common language of mediation—dwelling at the point of conjunction of Islam and Hinduism, Hindustan and Khurasan or Turan, medievalism and modernity, this world and the other. The fifth guru's special task was further to embody the scripture in the vernacular as the true centre, to reflect and be reflected in other true centres and to show moral discrimination of the false ones. He gave his life to distinguish the way of martyrdom *versus* kingdom, represented by the false centre at Lahore; and the ninth guru willingly did the same for toleration of the freedom of conscience at Delhi, capital of the empire. The tenth guru, through himself becoming the pupil of his pupils, as Bhai Gurdas II puts it, transferred the charisma and the task of the one tradition and the tradition of the one to the society for salvation, the pure and the elect (Khalsa), for the future which is also ours. At any rate, our investigation in these two chapters of the Muslim and the Sikh relation in history and theology casts this sort of a reflected light on the structure and unfolding of Sikhism itself.

(*c*) The project of the Indian modernity is thus to fold up in the name of God the inherited dualisms of collective and individual, the exoteric and the esoteric, status versus power; to find and disclose the true centre of self-rule, self-sacrifice and self-reform (*swaraj*); and to re-establish for the future equally in history and society as well as the self, through the method and praxis of non-violent action, 'martyrdom versus kingdom', the perennial philosophy of God's truth as non-dualism, unity in variety, in relation to self, the world and the other.

5

Martyrdom, Non-violence and Revolution

I believe God would make me an instrument of saving the religion that I love, cherish and practise.

Gandhi, *Harijan*, 1 February 1948

I shall have won if I am granted a death whereby I can demonstrate the strength of truth and non-violence.

Gandhi, letter of 24 January 1948, *CW* 90: 489

No religion in the world can live without self-suffering. A faith gains in strength only when people are willing to lay down their lives for it. The tree of life has to be watered with the blood of martyrs, who lay down their lives without killing their opponents or intending any harm to them. That is the root of Hinduism and of all other religions.

Gandhi, *Harijan*, 11 May 1947

My claim to Hinduism has been rejected by some, because I believe and advocate non-violence in its extreme form. They say that I am a Christian in disguise.

Gandhi, *Young India*, 29 May 1924

Non-violence and Gandhism

I have no more to say about Sikhism to the modern European view, liberal or Marxist, and what one has said so far can be readily re-stated in a formula of two lines. In terms of the language common to these interlocutors, the problem of the Indian modernity since 1500 is how to break through simultaneously or at one stroke the two intersecting dualisms already defined as forming the elementary structure of medievalism, status versus power and the individual versus the collective. And as part of the same or as the next step, how to produce the new morality of the community *and* the person, religion *and* politics or economics, that would answer the needs simultaneously of a multi-religious nation, a modern plural society and a federal secular state. The first row of relations or oppositions below expresses the manifest problematic of the Indian modernity; the second row expresses, as I have argued, the implicit contribution of Sikhism to it.

Multi-religious nation	:	modern plural society	:	federal secular state
		::		
UNITARIAN	:	VERNACULARIST	:	DEMOCRATIC.

It remains now to attempt an answer to our third interlocutor, the traditionalist view, which sees no need for radical self-reform in the Indian reconstruction of modernity nor for any deconstruction of the concept of tradition. It would attempt to claim Gandhi as the exemplar of Hindu tradition in the modern age, on the one hand, and also to reject him, on the other hand, in his relation of equality to non-Hindu traditions, incidentally declaring the former to be essential and the latter to be contingent to Hinduism. From the point of view adopted by his assassins on the morning after freedom and Partition, Gandhi was not at fault, however, merely in being tactically too obliging to the Muslims, for instance, but even more so because he had become infected in principle by some imported arrogance of the Christian martyr complex, since he was seeking to suspend the laws of ***karma*** through the novel element of vicarious suffering and redemption. He had mistakenly come to believe, as they appear to have inferred, like the so-called son of God *and* son of man, that if but one person could make oneself perfectly truthful and non-violent in thought, word and deed, then as a representative he or she would thereby save the whole race, religion or nation. The Sikhs were of course equally polluted with the same arrogant infection, perhaps via Islam, except that they believed, as a contribution of their own, that the saviour or the redeemer need be, not only one person, but rather a group of five or in its multiples, producing a collectivity to manifest God-man in history: the precedents for martyrdom versus kingdom as the best form of the marriage of religion and politics that Gandhi and the Sikhs tried to find in Indian history were contingent, opportunist or bogus (e.g. Baba Kharak Singh v. Master Tara Singh).

It is true, and apparently in favour of the traditionalist view, that Gandhi spent a good deal of his time in criticizing and indeed deploring the prevailing conditions of conversion from one traditional religious denomination to another in India. Nevertheless, as supreme witness to God's truth of unity in variety, specially Hindu and Muslim, Brahmin and Harijan, the same Gandhi spent his whole life himself trying to convert the world to the non-dualism of truth and non-violence, the perennial religion of India. He always argued that this was the one religion behind all religions, the moral law of our being and becoming, or that, at any rate, it could be held simultaneously with any existing one of the others, so helping to perfect rather than to weaken the latter.

In the political field, Gandhi thought that truthful, fearless and non-violent action (*satyagraha*), avoiding the slightest apparent intimidation or show of strength or force against the other, depended upon the *satyagrahis* advancing in the cause to engage themselves in single or at the most double file. The Sikhs, on the other hand, while protesting loudly their perfect adherence to non-violence in thought, word and deed during the *gurudwara* movement, 1920–25, felt that Gandhi had not sufficiently understood the role of the 'communion of saints' or the place of the congregation in the Sikh project. They wanted in action the 'procession of martyrs' (*shahidi jatha*) to advance in groups of five at a time or in multiples thereof, which I believe they eventually did—to the destruction of personal life or liberty and complete non-violent victory for the cause of reform—and incidentally to Gandhiji's entire ideological satisfaction.

This aspect of the matter seems to have again escaped the

attention of the latest student of Sikhism and Gandhism in the period of conjunction, 1920–25, which gave birth to the Akali party (Partha N. Mukherji 1984). I shall quote, therefore, more or less the same sources that he does, but as an experiment arranged in the reverse chronological order, in order to bring out, without any further word from me, some permanent aspects of this great recent example of the Indian modernity in both its method and praxis. It shows when and how to separate religion and politics (e.g. restoration of the Maharaja of Nabha, whose own morals were doubtful/doubted, and the *gurudwara* reform movement, whose morals had to be above reproach), and when and how to combine the two through making the common morality of self-rule the condition of self-reform, and equally self-reform the condition of self-rule (*swaraj*), at all levels, i.e. of the individual, the community, the party and the nation. One needs to listen to Gandhi's own words only, while bearing the modernist as well as the traditionalist views in mind, and so imagine the 'other' in his dialogue. The reverse chronological order adopted is suited as an experiment to ignore for the moment an historian's obsession with diachrony, cause and effect, and to see the structure of the 'universe of discourse' as a problem of synchrony.

Gurdwara Legislation, 11 July 1925

Both the Punjab Government and the Sikhs are to be congratulated upon the happy ending of the Akali movement. It has required the self-immolation of hundreds of the bravest in the land. It has required the imprisonment of thousands of brave Akalis. The public is familiar with the tale of their sufferings in the jails. Such marvellous sacrifice could not go in vain. Let us hope that the gurdwara reform will now proceed steadily and without a hitch. The Government deserve the congratulations, too, on their release of Akali prisoners and relaxation in the stringency of conditions regarding the *Akhand Path*. . . . If the conditions are not humiliating but merely precautionary or designed to save the prestige of the Government, I hope that the Akali friends will not raise unnecessary objection. Their chief aim was to attain the reform of the gurdwaras. This has been completely attained. The rest I regard as a matter of subsidiary, if not trivial, importance.

(Gandhi, *CW* 27: 361f)

The Akali Struggle, 26 June 1924

The public were hoping that the negotiations going on between the Akali leaders and the Punjab Government would bear fruit and that the Gurdwara question would be satisfactorily settled and the sufferings of the Akalis would end. But if the SGPC [Shiromani Gurdwara Prabandhak Committee] is to be relied upon, the Government had willed otherwise. The Akali leaders, it is stated, were all that could be wished, but the Government would not even promise to release those prisoners who are now undergoing imprisonment, not for violence actually committed or contemplated, but practically for having taken part in the Gurdwara agitation.

The Akali struggle will, therefore, in all probability be prosecuted with greater vigour. The Government will also probably adopt more repressive measures. Fortunately, we have now become inured to repression. It has ceased to terrify us. The Akalis have shown the stuff of which they are made.

Let us see what they have already suffered for what to them is a deeply religious question. I will say nothing of the Nankana tragedy, the [Golden Temple] Keys affair, the Guru-ka-Bagh brutality or the Jaito firing. I will not also say anything about S.G.P.C. being declared an unlawful association. The [Indian National] Congress has regarded it as a challenge to all public bodies that may be against the Government.

Since the Jaito firing the Akalis, recognizing [at Gandhi's instance] that their [collective] passive resistance to arrest was capable of being misunderstood for violence, have been regularly sending to Jaito Shahidi Jathas of 500 men generally every fortnight for quiet and submissive arrest. These allow themselves to be arrested without [making] any opposition whatsoever. They, on their arrest, are sent by special train to what is said to be a jungle and there detained without any trial, without any charge. Dry rations are provided which they have to cook for themselves. The jungle which is supposed to be malarial and overgrown with [elephant] grass passes muster for a prison. I understand that a few have died of fevers due to exposure and malarial climate. Over 3,000 prisoners are suffering in this fashion. Besides the Shahidi Jathas, smaller ones of 25 each have been crossing over to Jaito daily for the past nine months. They are taken to a station called Bawal and left there to shift for themselves. These Akalis often undergo severe hardships before reaching their destination. And so the awful routine goes on with clock-work regularity without apparently producing any impression on the authorities.

Why do these Jathas suffer thus? Simply for the sake of performing the *Akhand Path* ceremony which was rudely interrupted by the Nabha authorities and whose performance is even now being prevented. The Akalis have repeatedly stated that whilst they claim the right to demand and secure for the [deposed] Maharaja of Nabha an impartial and open inquiry, they do not want to use *Akhand Path* as a cloak to carry on any agitation in his favour. Why the *Akhand Path* is prevented no one can tell except that it is sought to crush the indomitable spirit of the Akalis which has organized and is carrying on the reform movement.

The demands of the Akalis seem to be absolutely simple. So far as I am aware, they are: (1) Possession of historical Gurdwaras by a central body elected by the Sikhs. (2) Right of every Sikh to possess a *kirpan* of any size. (3) Right of performing the *Akhand Path* in Jaito. On the face of it, every one of these demands is unexceptionable and should be recognized for the asking. No community has shown so much bravery, sacrifice and skill in the prosecution of its object as the Akalis. No community has maintained the passive spirit so admirably as they. Any other government but the Indian would long ago have recognized the demands and the sacrifice of the Akalis and turned them from opponents into its voluntary supporters. But the [British] Indian Government would not evoke the spirit of universal opposition which it has if it had cared for and respected public opinion.

The duty of the Hindu, Mussalman, and other sister communities is clear. They must help the reformers with their moral support and let the Government know unequivocally that, in the matters above named, the Akalis have the moral support of the whole of India.

I know that the distrust that pervades the Indian atmosphere has not left the Akalis free from the contagion. The Hindus, and possibly the Mussalmans, distrust their intentions. They regard their activity with suspicion. Ulterior motives and ambition for the establishment of Sikh Raj are imputed to them. The Akalis have disclaimed any such intention. As a matter of fact, no disclaimer is necessary, and none can prevent such an attempt being made in the future. A solemn declaration made by all the Sikhs can easily be thrown on the scrap-heap if ever their successors entertain any such unworthy ambition. The safety lies only in the determination of us all to work for the freedom of all. From a practical standpoint too, moral support of the reform movement, it is clear, reduces the chances of unworthy ambition being harboured in the Sikh breast. As a matter of fact, any such mutual suspicion necessarily hinders the *swaraj* [self-rule] movement for it prevents hearty co-operation between the communities and thus consolidates the forces of exploitation of this fair land and perhaps even makes possible an ambition which is at present clearly impossible. We must therefore judge each communal movement on its merits and give it ungrudging support, when it is in itself sound, and the means employed for its conduct are honourable, open and peaceful. (Gandhi, *CW* 24: 293ff)

The Gurudwara Movement, 17 April 1924

Another Jatha of 500 has surrendered peacefully when it was intercepted in its progress to the Gangsar Gurdwara and placed under arrest by the Nabha authorities. If we had not become used to such arrests and the like, they would create a sensation in the country. Now they have become ordinary occurrences and excite little curiosity and less surprise or pain. Their moral value increases in the same ratio as popular interest in them seems to have died. These arrests, when they cease to be sensational, also cease to afford intoxication. People who court arrest in the absence of excitement allow themselves to be arrested because of their unquenchable faith in [the] silent but certain efficacy of suffering undergone without resentment and in a just cause. The Sikhs have been conducting the Gurdwara movement by the satyagraha method now for the last four years. Their zeal is apparently undiminished in spite of the fact that most of their leaders are in jail. Their suffering has been intense. They have put up with beating, they have stood without retaliation shower[s] of bullets and hundreds have been imprisoned. Victory therefore is a matter only of time. An offensive has been threatened on behalf of the Government. They are imprisoning innocent men who are marching in pursuit of a religious duty. They have declared their [collective] associations illegal. One wonders what further steps they can take to frighten the brave Sikhs. The latter's answer to any offensive on the part of the Government is not difficult to guess. They will meet each progressive step in repression with equally progressive determination to do or die.

(Gandhi, *CW* 23: 456f)

Advice to Akalis, 9 March 1924

. . . In order that the method is and remains strictly non-violent throughout, it is not enough that there is absence of active violence, but it is necessary that there is not the slightest show of force.

It follows, therefore, that a large body of men cannot be deputed to assert the right of SGPC's possession, but one or at the most two men of undoubted integrity, spiritual force and humility may be deputed to assert the right. The result of this is likely to be the martyrdom of these pioneers. My [own or Christian?] conviction is that from that moment the possession of the Committee is assured, but it may so happen that martyrdom is postponed and intermediate stages such as pinpricks, serious assault or imprisonment might have to be suffered. In that case and in every case till actual control is secured, there must be a ceaseless stream of devotees in single or double file visiting the gurdwara in assertion of the right of the Committee.

. . . The idea underlying meek suffering is that ultimately it is bound to melt the stoniest heart. It further deprives disobedience of the slightest trace of violence either active or passive. (Gandhi, *CW* 23: 229–32)

Open Letter to Akalis, 25 February 1924

. . . But I would ask the Akali Sikhs not to send any more Jathas without further deliberation and consultation with those leaders outside the Sikh community who have hitherto been giving them advice. It would be well to stop and watch developments arising out of the tragedy. One of the telegrams received by me tells me that the Jatha was and remained throughout strictly non-violent. [*a*] You have, from the very commencement, claimed that your movement is perfectly non-violent and religious. I would like every one of us to understand all the implications of non-violence.

I am not unaware of the fact [*b*] that non-violence is not your final creed. It is, therefore, doubly incumbent upon you to guard against any violence in thought or word creeping in the movement. Over 25 years of [*c*] the practice of non-violence in the political field has shown me as clearly as daylight that, in every act of ours, we have to watch our thoughts and words in connection with the movement in which we may be engaged. [*d*] Non-violence is impossible without deep humility and the strictest regard for truth and, if such non-violence has been possible in connection with movements not termed religious, how much easier it should be with those like you who are conducting a strictly religious movement? (Gandhi, *CW* 23: 211)

Notes, 15 December 1921

[On the basis of the reports of Lala Lajpat Rai and Agha Safdar from Lahore]. There seems little doubt that the Sikhs have behaved with wonderful courage and restraint. When born fighters become non-violent, they exhibit courage of the highest order. The Sikhs have historical evidence of such exhibition in their midst [i.e. precedents]. They are now repeating their own history. (Gandhi, *CW* 22: 5–7)

Speech at Maharashtra Provincial Conference, Bassein, 7 May 1921

We need such heroes [of non-violence] not only among Sikhs, but among Hindus and Muslims as well. (Gandhi, *CW* 20: 68)

Sikh Awakening, 13 March 1921

One of their elders once told me that the Sikhs do not believe in *varnashrama*; there is no high or low among them; there is no untouchability; they look upon idol worship as a sin.

. . . Sikh temples are known as *gurudwaras*. The reformers [elsewhere described by Gandhi as "puritans" and "a great party of purists"] believe that the standards in these *gurudwaras* have become lax and the *mahants* are too often impious scoundrels [protected by the Government]. Some of the *gurudwaras* have historic importance. The reformers think that it would be best to take over [collective] control of all of them. This movement, which has been going on for a number of years now, seeks to introduce changes in them and to put their management under a committee. [It] . . . has become more rigorous after the starting of non-cooperation. Fifty or a hundred men of such groups go and take possession of a *gurudwara*. They claim that it is not their intention to take possession by force; they suffer violence themselves but do not use any. Nevertheless, a crowd of fifty or more men approaching a place in the way described is certainly a show of force, and naturally the keeper of the *gurudwara* would be intimidated by it.

Whether or not it is true that there is display of force in an action of this kind, the biggest *gurudwaras* have fallen into the hands of the Akali *Jathas*, the latter having lost 160 men in the process.

The majority of these died while taking possession of their most important *gurudwara*, known as Nankana Saheb. It is 40 miles from Lahore. . . . Naraindas was the *mahant* in control of this *gurudwara*. He is said to be given to sensuality. Though a professed *udasi* [renouncer], he had kept a mistress. His annual income was estimated at Rs 5 lakhs. The Akali *Dal* had its eye on this *gurudwara*. . . . It was the early morning of Sunday, February 20, when the Sikh *Dal* arrived [at Nankana].

Naraindas had been apprehending an attack on the *gurudwara* for some months. He had made preparations, had collected weapons and gunpowder. A number of rooms had been built forming a kind of fort with openings for directing gun-fire. The main door was [newly] rivetted with massive steel plates. Things were so arranged that a man who had gone in could not come out alive, and that, once the gate was closed, no one [else] could effect easy entry. The shrine stands nearly in the centre of the circle of rooms. The floor inside is of marble.

That Sunday, Lachhman Singh and his band entered the temple. It is said that they had gone only for *darshan* [worship]. They had no intention of taking possession that day.

Naraindas was already possessed with fear. A guilty mind is all timidity. He had become desperate. He looked upon the Akali *Dal* as his enemy. The moment Lachhman Singh bowed his head before the Granth Saheb, the assassins hired by Naraindas, who are stated to have taken up positions on the terrace of the cells, opened fire. I saw bullet marks on the Granth Saheb and on the columns of the marble cupola.

Lachhman Singh fell. It is said that, in that wounded, bleeding state, he was dragged, tied to a near-by tree, and burnt alive! I did see the burnt trunk of the tree and even the blood trails.

. . . But the *mahant* had run amuck. He had hired murderers with him. He decided to kill all the men. . . . He out–Dyered Dyer [of the Amritsar massacre, 1919]. Nor was this all. As if ashamed of his cruelty or wishing to hide the shameful fact that not a single man of his had been killed, this terrible *mahant* had the corpses sprayed with kerosene and burnt. Not a single man out of all those who had entered the temple could come out alive. Not even one witness on the Akali side has yet been found. . . .

In this way, for the sake and in the name of religion, over 150 Sikhs laid down their lives and established their title to the [SGPC] ownership of the *gurudwara*.

. . . On the day on which these martyrs were being cremated, we two, Maulana Abul Kalam Azad and I, were present at a meeting of Sikhs. He uttered one profound sentence there. "The blood," he said, "of a hundred and fifty martyrs has purified one *gurudwara*. Should it be any wonder if all of us have to be martyrs to purify the *gurudwara* that is India?"

Let us examine a little the sacrifice of these friends. [*a*] If they had intended to obtain control of the *gurudwara* by a show of force their aim was pure but their means should be considered impure. Since, however, they laid down their own lives, the world will always applaud their gallantry.

[*b*] If they had gone only for *darshan* [worship] and died [fighting] in self-defence, even then the world would admire their bravery and their means would not be called into question. [*c*] If, however, they had gone only for *darshan* and, notwithstanding that they were armed, they silently and without once raising their weapons embraced death, they are to be held to have given a demonstration of peaceful gallantry the

like of which no one has shown in modern times. If that is what happened, it could, in this age, happen only in India. What makes one so happy is that almost without exception every one of the Sikhs to whom I have talked on this subject believed that these 150 heroes had gone only for *darshan*, and that though it would have been possible for them to draw their swords, they refrained from doing so and perished, since they had taken a pledge to act peacefully.

If so, this is a perfect example of non-violent non-cooperation, and I firmly believe that its impact on the freedom movement will be tremendous. . . . We cannot imagine the limit to which we can increase our strength through self-immolation.

(Gandhi, *CW* 19: 397, 400, 422ff)

Message to Lahore Sikhs on Nankana Tragedy, 4 March 1921

The Akali party seem to have been treacherously admitted and the gates closed on them. Everything I saw and heard points to a second edition of Dyerism more barbarous, more calculated and more fiendish than the Dyerism at Jallianwala [Amritsar]. Man at Nankana, where once a snake is reported to have innocently spread its hood to shade the lamb-like Guru [Nanak], turned Satan on that Black Sunday. India weeps today over the awful tragedy. I am ashamed to find that there are men today who are capable of the crime committed by sons of India in that holy temple.

. . . I can only think of the tragedy in terms of Indian nationality. The merit of the brave deed must belong not merely to the Sikhs but to the whole nation. And my advice, therefore, must be to ask the Sikh friends to shape their future conduct in accordance with the need of the nation. The purest way of seeking justice against the murderers is not to seek it. The perpetrators, whether they are Sikhs, Pathans or Hindus, are our countrymen. Their punishment cannot recall the dead to life. I would ask those whose hearts are lacerated to forgive them, not out of their weakness—for they are able in every way to have them punished—but out of their immeasurable strength. Only the strong can forgive. You will add to the glory of the martyrdom of the dear ones by refusing to take revenge.

(Gandhi, *CW* 19: 399ff)

Speech at Nankana Saheb, 3 March 1921

. . . It seems almost unbelievable that not a man died at the hands of the Akali party. Did not the brave men who were armed with ***kirpans*** and battle-axes retaliate even in self-defence? . . . I hope that you will not take the credit of the bravery for the Sikhs only, but that you will regard it as an act of national bravery. The martyrs have died not to save their own faith merely but to save all religions from impurity.

. . . I would ask you therefore to dedicate this martyrdom to Bharat Mata [Mother India] and believe that the Khalsa can remain free only in a free India. You cannot be free and enslave India. . . . Your *kirpans* must therefore remain scrupulously sheathed and the hatchets buried. If you and I will prove worthy of the martyrs, we will learn the lesson of humility and suffering from them; and you will dedicate all your matchless bravery to the service of the country and her redemption.

(Gandhi, *CW* 19: 397f)

Martyrdom and Revolution

[*a*] In *Of Prelatical Episcopacy* . . . Milton was less interested in the fact of suffering than in the truth of the position taken. This emphasis was consistent with [Saint] Augustine's principle that the cause, not the punishment [or suffering], makes a martyr, *Martyrem non facit poena, sed causa* (*Epist.* 89: 2).

[*b*] When Milton took up his commission to answer [it] . . . for him *Eikon Basilike* presented not the truth but a carefully wrought image of a Charles [I] who discredited his claims to martyrdom by his self-conscious preoccupation with the wrongs done him: *Martyrs bear witness to the truth, not to themselves. If I beare witness of my self, saith Christ, my witness is not true* (*Eikonoklastes*, 3: 575).

[*c*] In *Samson Agonistes* the collapse of the temple affords spectacular confirmation of the power of the *living God* that Samson serves and validates his role as God's champion. . . . Milton's usage here underscores a kind of reciprocity that seems to have appealed to him, whereby the testimony of the faithful is answered by a divine response that vindicates it (Knott 1993: 155, 160, 172).

To continue with our experiment to determine and define the elementary structure of martyrdom, I shall now remove the technical requirement of non-violence as a creed in terms of means, but the other moral requirements of truthfulness, fearlessness, poverty and chastity for the martyr as *satyagrahi* will remain as before. It is advisable, moreover, to retain the condition or the context of situation as one of internal conflict, as in the Indian case just examined, rather than one of foreign war, subjugation or persecution, so as not to be distracted from our study of the martyr to that of the hero and/or the victim. The martyr is one who must love his enemy in some sense since he or she is the perfect witness (*shahid-ul kamil*) that God, who at this time takes an interest in history and politics, does not want his servant to suppose, as the dualist would, that Satanism has any truly independent existence, and so *dharmayuddha*, the righteous war, can be transformed into *satyagraha*.

We again choose to give the Orientalist and the Occidentalist the go by (Wensinck 1922 and Knott 1993) and simply present the argument and evidence, here structurally arranged, adduced in his two lectures of 1972 by Ali Shariati (b. 1933), an original figure, who is stated to have held two doctorates from Paris in sociology and the history of religions. He and Bani Sadr were perhaps the two chief ideologues of the Islamic revolution of Iran. It is supposed that Shariati was obliged to flee his homeland upon discovery of his secret nightly lectures in 'other people's homes', some five years later, but was tracked down and assassinated in London by agents of the Shah of Iran's secret police (d. 1977).

As a Shiah Muslim, Shariati no doubt preferred the theory of the imamate to that of the caliphate, the hereditary principle to the elective principle, being much more concerned with inheritance and transmission of the divine light in the family of the Prophet, 'this little house of Fatimah which is greater than all of history'. Nevertheless, he evidently supposes that the imamate, idealized like the caliphate, 'consists of a grave mission of guiding and driving a society and individual, from "that which is" [fact] to "that which should be" [value] . . . based upon a permanent ideology which the Imam himself obeys' (N.D.: 74, 115).

In all ages and centuries, [*a*] when the followers of a faith and an ideology have power, they guarantee their honor and their life with *jihad* [effort in the way of God]. But [*b*] when they have fallen into weakness and all means of struggle have been taken away from them, they guarantee their lives, movements, faith, respect, honor, future and history with martyrdom, for [*c*] martyrdom is an invitation to all ages and to all generations, that if you can, kill, and if you cannot, die.

At first sight this statement makes out martyrdom to be chiefly a means forced upon one by circumstances, and so reduces the status of the martyr to the status of the hero, but the juxtaposed paragraph gives the better statement of martyrdom as both a means and an end, which I take to be Shariati's view and intention.

Martyrdom, in summary, as opposed to other histories, where it is an accident, an event and a death imposed upon a hero, a tragedy, in our culture, it is a grade, a rank, a level. It is not a means but a goal unto itself. It is genuine. It is a transformation, an elevation. It is itself a great responsibility. It is itself a short-cut towards rising to the summit of the ascension of humanity and it is a culture (N.D.: 79).

The difference between the two statements is manifest in detail in Shariati's retelling of the well-known story of Husayn, second son of Ali and Fatimah and grandson of the Prophet, whose martyrdom is the central event of Twelve–Imam Shiah history, as seen every year during the tenth day of Muharram in India and elsewhere. 'For the Shi'ites, Husayn is an Imam, a figure they understand to be an intermediary between man and God; they believe this function to be hereditary, inherited from 'Ali and passed on to Husayn's descendants' (Glassé 1989: 162). Shariati takes very great care to establish, against all error and calumny, that Husayn's uprising was freely intended to preserve the cry of the Islamic revolution through his own death—and his sister's message of remembrance of it. Husayn knew well that the odds against his practical military or political success were impossible, he had been amply forewarned against his course of action, but he simply had no wish to win the struggle for power and assume the caliphate to which he was otherwise entitled against the usurper Yazid, who had in his favour neither birth nor election. Shariati is therefore arguing against the opinion which says, 'The Imam arose officially and armed to destroy the Umayyid regime and to take over their rule and leadership of the people', and that he was defeated by the treason of the people who did not come to his aid, the tricks of the enemy and the *coup d'etat* in Kufah, that is, 'He arose against the enemy to attain victory, and not to be killed' (N.D.: 83).

Firstly, there is to be considered the 'definition of the

situation', as sociologists call it, when Islam was in danger of becoming an instrument to justify the regime and whatever it did, and the swords of *jihad* were being put to use by executioners under the usurping power.

> [*a*] Companions [of the Prophet] who were loyal, persisting *mujahid* [one engaged in *jihad*] have been killed and silenced. [*b*] The chaste Companions who do not sell out, prefer a safe retreat and ritual devotions over the headache of truth and falsehood [usurpation and non-usurpation] and the dangers of a political and social struggle. They released themselves from people and their fate and crawled into their shells of respectful, ascetic self-worship and kept mum. [*c*] Another group of the most notable Companions of the Holy Prophet spend their time around the Green Palace of Mu'awiyyah, [with] their heads in the troughs of the public treasury. [*d*] The young generation, the second generation of the [Islamic] movement, in their endeavors and in their resistance struggles against the Bani Umayyid, [have] experienced defeat. [*e*] With the sword, money, position or cunning, the ruling power has calmed, stifled and silenced everyone (N.D.: 36f, 41, 45, 47).

Secondly, nevertheless, the holy war for the hearts and the minds of men must proceed if Islam is not to return to the pre-Islamic 'age of ignorance', new idol worship, slave merchants and the capitalism of the Quraysh. Otherwise,

> Little by little among the intellectuals of society, questions concerning the soul, the body, prime matter, accidents, love and I don't know . . . the first station and the last station, attraction and ecstasy, etc. came up but issues like responsibility, people, truth, leadership, and . . . have all been forgotten.
>
> . . . The principle is that when the Imam is confronted by a usurping government, he must undertake an armed political uprising, remove the powerful, reactionary regime through revolutionary power, revive the truth and undertake the leadership of the people. . . . This is also accepted but Imam Husayn's movement is not the motion of a political or military upriser . . . [though] not because—as some say— . . . it is below his dignity to turn to politics and a political revolution. No. This is a duty. . . . Then why should the Imam's arising not be a political or military arising against the Umayyid regime? (N.D.: 47–50).

To find the answer, *thirdly*, Shariati will consider the 'form of the movement' when Husayn leaves Medina and goes to Mecca, following the same route between the two cities as did the original migration of the Prophet, but now traversing it in the opposite direction.

Then he arrives in Makkah. He openly travels the 600 kilometres from Madinah to Makkah accompanied by his family and there, also, he announces to all of the pilgrims who had come from all of the Islamic lands to perform the *hajj*, "I am going towards death."

A person who is undertaking a political uprising does not speak as this. He would say, "We will fight. We will kill. We will be victorious. We will destroy the enemy," but Imam Husayn, in his movement and in his address to the people, says, "Death for the children of Adam is as beautiful as a necklace around the neck of a young and beautiful girl."

. . . He announces it to the regime, to the authorities, to the army, and to all of the ruling power, to the people, officially, openly, decisively and clearly, "I refuse to give my allegiance [to Yazid]. I will go from here. I have undertaken a migration to death" (N.D.: 51ff).

Fourthly, what precisely is the danger, then, facing both religion and humanity that Husayn still has the duty to fight in some way when he has certainly the status but clearly not the power to do so against Yazid and his men, who have the power but not the virtue of Islam, and how may the choice of martyrdom resolve this contradiction? 'All of the custodians of intelligence and religion, advisors of Divine Law and common law, expediator-seekers [realists] of goodness and logic, all unanimously say, "No!", but Husayn [alone] wants to say, "Yea!" ': he must be responsible for and 'loyal to the truth and justice and the freedom of Islam'.

It is an age when there is no longer a cry from among the people. . . . And now, Husayn . . . sees that if he remains silent, all of Islam will take the form of a government religion. Islam will be transformed into a military-political power and nothing more—an event and then a regime, a ruling power like the power of the Mongols, like the power of the government of the Seljuks, or the successors of Alexander, or the successors of Ghengis Khan and then, when the power itself is destroyed and when the caliphate and the army have been dispersed and ruined, nothing will remain but a memory in history, an event of the past which is over and done with.

This is why Husayn now stands before two inabilities. He cannot remain silent because . . . whatever was delivered by the message of Muhammad and Islam through efforts of *jihads* and sufferings will be obliterated. . . . On the other hand, he cannot fight because he does not have the power to fight [a war]. He is in the strong clutches of the ruling power. Neither can he cry out nor can he remain silent. Neither can he surrender nor can he attack. . . . But the heavy burden of all of these responsibilities rests solely upon his shoulders. . . . In this school, a person who is alone is also responsible to oppose oppressive absolute rule which determines the fate of the people, because responsibility is born from awareness [of freedom, or being human as against being Yazid] and faith, not from power and possibility. . . .

In the midst of this, only one man—a lone man—says, "Yea!" What does that "yea" mean? It means that in the midst of absolute inability, absolute weakness, in an age of darkness and silence, an aware, liberated, man of faith, who still has the responsibility of the *jihad* against usurpation and oppression, Imam Husayn's edict is this: "Yea." In "inability" there rests a "must". For him, life is ideology and *jihad*. . . . "Able" or "unable", weakness or strength, alone or collectively, [can] only determine the external form of the performance of the mission, the "how" of the realization of responsibility, not its existence. . . .

The great teacher of martyrdom has now arisen to teach those who consider *jihad* to be only in "ability" and to those who see victory over the enemy only in triumph that martyrdom is not a defeat, a loss, but that it is a choice, a choice whereby the *mujahid*, by sacrificing himself at the threshold of freedom and the *mihrab* [altar niche in the mosque] of love is victorious. . . . The heir of the great prophets who had taught humanity "how to live", has now come in this age to teach the children of Adam "how to die" (N.D.: 53–7, 69f, 75).

Fifthly, we seem to approach the conclusion somehow (*a*) that all martyrs so defined, like the Christian saints, are intermediaries between man and God, in the sense that one prays to them for intercession and not for them; and (*b*) that this position lends to the free act of martyrdom the meaning and effect of representation or vicariousness in relation to the suffering and redemption of others. At any rate, Shariati is quite clear that, for example, when Husayn and his company willingly courted martyrdom, they truly bore witness to the benefit of all the oppressed of history with their blood, not only with words, that martyrdom is the greatest potentiality placed at the disposal of 'other than God'.

The word *shahid* (martyr) contains the greatest sense of meaning of that which I have tried to say and want to say. *Shahid*, in definition, [*a*] means "being present", "observer", in the sense of "a person who bears witness" and is a "trustworthy person who speaks about what is real and true" and [*b*] it also means "aware" as well as "sensible" and "perceptible", a person who everyone looks up to and, finally, [*c*] it means, "model, pattern, example". . . . In other religions and in the histories of nations, martyrdom is the sacrificing of the heroes who are killed in battle by the enemy. It is considered to be a tragic event, full of sorrow and the name given to those killed is martyr, and their death is called martyrdom. But in our culture, ***shahadat*** (martyrdom) is not a death which our enemy imposes upon a ***mujahid. Shahadat*** is a desired death which a ***mujahid***, with all of his consciousness, logic, reasoning, awareness and understanding chooses himself.

Husayn . . . arises . . . to disgrace the enemy and to tear asunder the curtains of deceit. . . . And he has nothing for *"jihad"* other than his "being". He takes that and leaves his home to move towards his place of martyrdom . . . with a unique selection of his Companions—men who had come with him to die—and also with each and every member of his family . . . to sacrifice all of them at the threshold of martyrdom . . . [so that] he bears witness: "In an age when the truth was unarmed and defenseless, I carried out my responsibility; bear witness that I could do nothing more than this." This is why you have heard that on 'Ashura [the day on which Husayn and 72 of his household and companions were massacred in the desert at Karbala], he takes the blood flowing from the throat of his child in his fist and throws it towards heaven, saying to God, "Look! Accept this sacrifice from me. Witness!" (N.D.: 75ff, 101, 120).

I infer from this narration that, for any complete definition, (*a*) the twofold relation has to be evoked that, just as the martyr bears witness to the truth of faith in his age before God and humankind through self-immolation, so reciprocally will God and humankind be witness to him for all time and eternity; and (*b*) that the notions of history and society, in which we are all members of one another, are equally essential to the structure of martyrdom, which is the total gift of self as the highest form of love in relation to the state of the world and the other. Thus, on the one hand, for himself, 'martyrdom consists of an action whereby a man suddenly and in a revolutionary manner, throws his abased and inferior being into a fire of love and faith, becoming total light and sacred'; while for others, and perhaps for God, on the other hand, 'a martyr is the heart of history . . . and the greatest miracle of his martyrdom, arising to bear witness, is to give a generation new faith in themselves', which makes his or her self-immolation different from suicide (N.D.: 78, 105).

Sixthly, one can now summarize Shariati's theory of martyrdom as a product of the twentieth century, and also ask where it belongs in the logical economy of Shiah Islam. Its elementary structure is manifestly the non-dualism of loving self-sacrifice, as we have just shown, but equally it is the responsibility of 'arising to bear witness' in the duality of the true and the false, religion and irreligion, liberation and bondage. Martyrdom is to be 'present at the scene' in Shariati's terms of the eternal battlefield of truth versus falsehood in history; and its moment has come when the worshippers of truth require self-sacrifice in society.

Martyrdom is that which they [false usurpers and tyrants] desire to have disappear from history. It is [both] a model of that which should be, [and] it bears witness to that which takes place in the silence and secrecy of time.

And, finally, it is the only form of *jihad*; the only reason for existence; the only sign of being present; the only weapon of attack and defense; and the only method by which truth, right and justice can resist in an age and under a regime in which falsehood, deceit and suppression [prevail and] . . . Humanness has found itself at the threshold of decline, facing the danger of dying forever.

He has no weapon other than death . . . [yet he is confident of victory]. In this world, there is no one who knows as he knows how one should die, a knowledge which his powerful enemy, who is ruling the world, is deprived of.

He [Husayn] has died in Karbala in order to be resurrected in all [future] generations, in all ages.

"They can live well who can die well" . . . because everyone dies the way he has lived.

It is because of this that [in Islamic law] a martyr [*a*] does not need to have the ritual ablution performed [after death], [*b*] has no shroud. A martyr [*c*] has no accounting on the Day of Judgment [and does not need the Prophet's intercession], because [*d*] the person who had sinned and erred has been sacrificed before death by the martyr and now finds [God's] presence (N.D.: 32, 74–8, 106f).

Moojan Momen, writing on Shiah Islam in general, explains

to us the special role of Husayn as martyr in the family of the Prophet. He does not try to explain, as I have done, how the two well-recognized modes of pious Shiah existence, namely, meek patient suffering (*mazlumiyyat*) and the rising up (*qiyamat*), are perfectly reconciled in the elementary structure of martyrdom as 'arising to bear witness in loving self-sacrifice'.

Muhammad is, of course, the recipient of the revelation, the link with God; he is, however, so exalted as to be only approachable through one or other members of the family; 'Ali represents the intellectual, esoteric side of religion (the way to obtain the true meaning of the revelation) and its legalistic aspect ('Ali had complete knowledge of the religious law and was the perfect judge); . . . Husayn represents atonement, his redemptive martyrdom gives to all the possibility of salvation; the Twelfth Imam is the focus of eschatological hopes of triumph over tyranny and injustice and final salvation. While the ulama look to the image of 'Ali, the image of the intellectual, esoteric yet legalistic [exoteric] attitude towards religion, it is undoubtedly Husayn and his representation of redemption through sacrifice and martyrdom that has caught the imagination and devotion of the Shi'i masses (Momen 1985: 235f).

Two further questions still remain to be addressed to Shariati, one concerning the connection of martyrdom with revolution, and the other concerning the status of women. Therefore, *seventhly*, we ask whether the relation of martyrdom and revolution is a categorical imperative or contingent upon circumstance? The answer of Shariati will depend upon the distinction that he makes between two classes of prophets and religious movements throughout human history based upon their respective 'social and class affiliation', roughly translatable as the aristocratic and the democratic. On the one hand, in the tradition of Abraham, we have,

a series of prophets whose common aspect, from the social point of view, is that they have all arisen from among the most deprived groups of the social and economic life of their time and with the confirmation of the Prophet of Islam himself, "They were all shepherds". . . .

It is because of this that the group which gathered around the Prophet in Makkah, other than a few, were from among the most deprived, the most forgotten of the scorned and the most scorned elements of society; and it is for this reason that the enemies of Islam scorned the Prophet of Islam: "Only the trash are around him", and this is the greatest praise today for this movement. . . .

But [on the other hand] the prophets outside this series or the leaders of intellectual and/or ethical schools of thought outside of this series [Confucius, Lao Tse, Buddha, Zoroaster, Mani, Mazdak, Socrates, Plato and Aristotle], whether they were in China, relating to the yellow race, or whether in India or whether in Iran or they were the leaders and founders of the intellectual or moral schools of Greece, without exception, they were all aristocrats and had arisen from the well-off, successful and powerful classes. . . . [The princes and the priests] whether they were concordant with each other or in opposition, at any rate, their agreement or lack of agreement was over who should rule over the people, not for the people.

The appointment of those prophets (non-Abrahamic) is most often connected to the existing power for the dissemination of their religion . . . whereas the actualization of the Abrahamic prophets was continuously in connection with the people against the existing power of their time.

Everywhere in the discussions of the Abrahamic prophets, the discussion is about the masses, the discussion is about the common people . . . [and] speaking in the language of the people, based upon special needs and understanding of the masses of the people; that is, not speaking in the language of the elite (N.D.: 17–22).

The internalization of this distinction, the aristocracy of privilege versus the democracy of shepherds, as a polarity or duality within the Muslim world itself is what forges the link between martyrdom and revolution for the Khariji, the Sufi and the Shiah alike; and we have already considered the consequent dialectics of secession, separatism and schism in chapter 2. It is precisely this process and this linkage which is everywhere resisted by the fatalists and the conservative ulema, who justified the caliphate and the sultanate on grounds of the necessities of time and circumstance. They always unitedly taught that rebellion against authority, howsoever it was constituted, was not a *jihad* with any prospects of martyrdom.

Lastly or *eighthly*, it turns out fortunately for us 'tongues which speak' that we were wrong to suppose that time and the Islamic revolution awaited but one person in 60 AH: there were in fact two, a man and a woman, and Shariati, who now lies buried next to Zaynab's tomb in Damascus, explains what their respective roles had to be, concluding with some lines of verse that may be the only ones that he ever composed. Imam Husayn had arisen in order:

To bear witness with his [sister] Zaynab that in the system which rules history, women had to either [*a*] choose bondage and being toys of the harems or [*b*] if they were to remain liberated, they must become the leader[s] of the caravan of the captives and the survivors of the martyrs.

Once again [as with Ali], from the silent and sorrowful house of Fatimah [daughter of the Prophet by his first wife, Khadijah], this little house which is greater than all of history, a man emerges—angry, determined . . . an illuminated and powerful face . . . heir of the great anguish of the human being, the only heir of Adam, the only heir of Abraham, the only heir of Muhammad. . . . But no! Walking side by side him, a woman has also emerged from Fatimah's house, step by step with him; she has taken half of the heavy mission of her brother upon her own shoulders.

Every Revolution has two visages: the first visage, blood; the second visage, message. *Shahid* means "being present". A person who chooses to select a red death as a means of showing his love for a truth and as the only weapon for *jihad* for the great values . . . they are alive; they are present; they are witnesses and they are observers, not only upon the threshold of God, but also upon the threshold of the masses and in every age, century, time and place.

. . . A martyr is showing and teaching and giving a message to [*a*] you who think inability excuses you from the *jihad*, that when confronted by oppression and suppression, [*b*] you who say victory over the enemy is realized when you triumph, no. [*c*] A martyr is a person who in a time of inability and lack [of political and economic means] of victory, is triumphant over the enemy with his death.

. . . Husayn and his Companions today passed through the first mission; the mission of blood.

The second mission is that of the message, [it] is to deliver the message of martyrdom to the ear of the world, is to be a tongue which speaks. . . . The mission of the message begins this afternoon. This mission is upon the delicate shoulders of a woman—Zaynab—a woman who [is such that] alongside her, manliness has learned manliness. And Zaynab's mission is more difficult, more grave [even] than the mission of her brother (N.D.: 73f, 102–10).

> Whoever knows
> What the responsibility of being a Shi'ite is . . .
> Should know

That in history's everlasting battlefield,
In the eternity of time
And in all places upon the earth:
All scenes are Karbala,
All months are Muharram,
All days are 'Ashura,
And
Must choose:
Either blood or the message;
. . . Those who left
Committed a Husaynic act;
Those who remained
Must perform a Zaynabic act;
Otherwise, they are Yazidic.

6

Texts and Rites of the Daily Divine Service

> Nonconformist worship had been restricted [to the non-congregational] since the collapse of Puritan power and the restoration of Charles II in 1660 . . . under the series of new Acts designed to enforce [congregational] worship according to the Book of Common Prayer and to prohibit other forms of religious assembly. . . . The Conventicle Acts of 1664 and 1670 subjected to severe penalties those meeting for unauthorized worship in groups of more than four [i.e. five or more].
>
> Knott 1993: 181

We may conclude from our experiment in the last chapter that, in theory and practice, martyrdom as a creed (*shahadat*) is more or less equivalent to non-violence as a creed (*satyagraha*) in the making of the Indian modernity from the birth of Sikhism up to Gandhism. The elementary form of martyrdom, as of Gandhian non-violence in the sense of *ahimsa*, is that of 'arising to bear witness in loving self-sacrifice to truth', which cancels the duality of virtue and power, the exoteric and the esoteric, individual and collective. As we have defined it, it is a form of the non-dualism of self, the world and the other, and it must be expressed in the common morality of religion, family, economics *and* politics, which is the Indian modernity of history and civil society.

I may now add that a study of Sikhism also makes explicit what had always remained occult in Shiah Islam, namely, that the relation of martyrdom and revolution implicitly replaces the concept of tradition with the concept of society, the common usage and custom, although the Sunnis admit of no conflict and always refer to themselves as *ahl-i sunnat wa jama'at*, 'people of tradition and the community', even after AD 1200 when the role of consensus (*ijma'*, lit. assembly) had petered out completely. On the other side, it is a fact in favour of my line of interpretation that, alone among those who were close to the origin of Islam, no usage, precedent or tradition of the Prophet is transmitted or handed down through Husayn, the martyr.

In the mode of thought and code of conduct of Sikhism, the other half of God and the scripture, the recorded and received gurus' word and example, is solely the congregation of the faithful, the communion of saints and martyrs, the community, special (*samuh*) and general (*sarbat*), that carries the name of the Khalsa, the pure and the elect, ever since 1699. The word and the work of Guru Arjun, the fifth guru, in the first decade of that century, is I think definitive in both these regards in fact and value, theory and practice, and we may now examine the elementary structure of modernity from this point of view in the internal rather than the external relations of Sikhism. At any event, the Orientalist who is familiar with the Punjabi sources agrees with me that 'The divine Guru, present within each of the ten personal Gurus, now dwells eternally in the corporate community (the Guru Panth) and in the sacred scripture (the Guru Granth)'; and that this represents the 'ultimate form of the doctrine' of Sikhism defined as the trinity of Guru, Granth and Panth (McLeod 1984: 78).

The system of Sikhism as method and praxis is reducible, without too great exaggeration at the end of this study, to three aspects or principles, which together stake its claim to the freedom and the responsibility of self-rule and self-reform, whether one calls that the Indian modernity or the eternal religion of truth, love, justice and self-sacrifice, or both. (*a*) The cult of the Name or the Word, which attempts to reconcile the esoteric and the exoteric, individual and collective, in the form of worship using the vernacular ever since AD 1500. (*b*) The cult of remaining forever unshorn in the world, in the state otherwise commonly known as *anand*, in love and fear of a God who, perhaps in return but certainly in his grace, takes an interest in history, economics and politics. (*c*) The cult of divine service (*seva*) specially within a society for self-realization, the Khalsa, special and general or singular and plural, the symbol of which is serene non-violent martyrdom.

I would not be understood to say that the trinity of Guru, Granth and Panth forms a simple unity of identity, but that there subsists a more complex dialectical unity of opposition and mediation such that, if and when any two of the terms come to be mutually opposed, it invokes the mediation of the third term. The invisible Guru and the visible Granth, if they are contrasted, call forth the Panth that is both. The personal Guru and the group of the Khalsa are reconciled in the Granth's body, the-scripture-in-the-congregation. Where the Panth is in doubt and the mere words of the Granth seem not to answer, then the Guru will speak through the spirit and the letter of the two together, as happens when the Guru Granth is consulted by the congregation. Indeed, everyday liturgy, discussed in this chapter, as well as events of crisis in the life of the individual and the community are marked by just this trinity of relations.

Guru-Granth

The Guru is in fact God, as McLeod explains from the sources, the Creator himself acting as Preceptor in order to reveal the message of salvation to humankind.

According to the Adi Granth God is a God of grace who actively communicates the truth which sets men free. The means whereby he effects this communication, the "Word" of God, is mystically "spoken" to man's inner understanding by the Guru. The Guru is thus, in a primary sense, the "voice" of God and it is in this sense that it is used [below] in the refrain from *Sri Raga* 1. Because Nanak submitted himself to God in total [active] obedience and achieved thereby a perfect enlightenment he became the human vehicle of the "voice" of God. As such he is properly designated Guru Nanak. The same divine spirit similarly inhabited his nine successors, all of them bearing the same exalted title of Guru or Satguru (True Guru). With the death [or the initiation] of Guru Gobind Singh the divine Guru ceased to occupy an [individual] human habitation and passed instead to the sacred scripture (the Guru Granth) and the company of believers (the Guru Panth).

After its opening liturgical prologue, which includes the *Japji*, the Adi Granth takes up its structural sequence of compositions in thirty-one *ragas* (metres). The first *raga* is *Sri raga*, and since the works of the gurus are recorded in chronological order within each *raga*, we begin with a hymn of the first guru. 'In this opening hymn,' the Orientalist says, 'Guru Nanak enunciates his fundamental doctrine.'

If I should own a priceless palace, walled with pearl and tiled with jewels; / Rooms perfumed with musk and saffron, sweet with fragrant sandalwood; / *Yet may your Name remain, O Master, in my thoughts and in my heart.*

[*Rahau*, pause] Apart from God my soul must burn, / Apart from God no place to turn, / The Guru thus declares.

[*a*] If, in a world aglow with diamonds, rubies deck my [red marital] bed; / If with alluring voice and gesture jewelled maidens proffer charms; / *Yet may your Name remain, O Master, in my thoughts and in my heart.*

[*b*] If with the [*siddh*] yogi's mystic art I work impressive [*siddhi*] deeds / Present now, then presto vanished, winning vast renown; / *Yet may your Name remain, O Master, in my thoughts and in my heart.*

[*c*] If as the lord [*sultan*] of powerful armies, if as a king enthroned, / Though my commands [*hukam*] bring prompt obedience, yet would my strength be vain. / *Grant that your Name remain, O Master, in my thoughts and in my heart* (transl. McLeod 1984: 38ff).

The message to be gathered from such important texts, as it appears to me, is that the discipline of the name and the religion of self-sacrifice in the society for salvation is not to be realized either by possessing or by renouncing the qualities of any of the three spheres of (*a*) *grihasth*, (*b*) *sannyas* and (*c*) *rajya*, as indicated in the three consecutive paragraphs above, but through finding their point of conjunction or the state of being forever unshorn (*anand*), by its new social definition. This message of the scripture certainly begins with the first guru, then, and confirms the view taken from the side of the community's five symbols, for example, with which we began in chapter 1. Here I simply continue the exposition with another example from a modern work of commentary on the nature of the Sikh religion, the name and the guru, namely, Bhai Jodh Singh's *Gurmati niranay*, 1932.

From these examples it is evident that in Sikh usage the word *guru* assumes those meanings which were originally associated with *avatar*. As such it refers to those special beings [e.g. prophets] sent by God to reveal the path of truth again. It does not possess the same meaning as ordinary Hindus attached to the word *guru* prior to the coming of the Sikh faith.

Before the Guru's teachings were delivered it was believed in India that whenever ignorance spread and men disobeyed their sacred duty God assumed the form of an incarnation (*avatar*) to restore men to the path of truth. According to *Gurmat* God is never incarnated. The Sikh belief is that at such a time it is the Guru who appears. . . . In the *janam-sakhi* account . . . it is clearly stated that Guru Nanak received the "cup of the Name" directly from God himself. The tenth Guru has also declared: "Know that the [only] eternal and incarnate One is my Guru".

Mercifully the True Guru has himself enumerated in the sacred scriptures the criteria which enable the ordinary person to recognise the Guru. The first of these is [*a*] that the Guru who would lead the disciple to union with God must himself have attained that perfect union. . . . [*b*] He must also be one who preaches his message and seeks to lead others to the divine Name without any concern for his own gain. . . . And [*c*] he must be free from all enmity. These are the qualities which enable a disciple to recognise the Guru as a true representation of God. And his words are God's words.

It is by means of the Word alone that the disciple is made one with the Guru. . . . And so the scripture which incorporates the utterances of the succession of personal Gurus is in fact the Guru.

The practice of "repeating a name" is also enjoined [as an invocation]. . . . But which name should be repeated? The Gurus have given us many names which illuminate the nature of God. There is, however, one particular name which [although it occurs later than the first guru] is widely used in Sikh *sangats* [congregations] and which has been commended by those such as Bhai Gurdas [*fl.* 1600, amanuensis of Guru Arjun and the first theologian of Sikhism] who have described the practice. This is "Vahiguru". The Tenth Guru himself invoked this particular name when he promulgated the greeting: *Vahiguru ji ka Khalsa, Sri Vahiguru ji ki fateh!* [God's is the Khalsa; may God's be the victory!] (transl. McLeod 1984: 139ff).

Jodh Singh and McLeod both fail to notice, however, that all the commonly favoured names, Sat-nam (the true Name), Vahiguru (the wonderful Lord) or Akal Purakh (the eternal One), have also an implicit diacritical function, since they are to be spoken loud and clear, indeed full-throated in greeting and ceremony, and are not to be held secret and silent, unspoken or merely whispered, as is the great name of God in some other tradition(s). The name *Vahiguru* is most commonly used as the term of address for the Almighty, for example, in the twice daily liturgy or congregational prayer of supplication (*Ardas*), morning and evening, as we shall see. To continue here our exposition of the-guru-in-the-scripture, we now ask how its message reconciles the individual and the collective or society and the person within the quest for salvation.

On the divine name, Guru Arjun's *Sukhmani*, 'pearl of peace' and/or 'peace of mind', a long poem of 2,000 lines, is probably the longest single writing in the scripture. Although its recitation is not an obligatory part of the daily individual divine service (*Nit-nem*), many Sikhs and non-Sikhs in fact include it as the centre-piece of early morning devotions.

So many scriptures, I have searched them all. / None can compare to the priceless Name of God. / . . . Better by far than any other way is the act of repeating the perfect Name of God. / Better by far than any other rite is the cleansing of one's heart in [and through] the company of the devout. / Better by far than any other skill is endlessly to utter the wondrous Name of God. / Better by far than any sacred text is hearing and repeating the praises of the Lord. / Better by far than any other place is the heart wherein abides that most precious Name of God. /

. . . God is infinite, beyond all comprehending, / Yet he who repeats the Name will find himself set free. / Hear me, my friend, for I long to hear / The tale which is told in the company of the saved. /

. . . Faces shine in the company of the faithful; / There, in their midst, sin's filth is washed away. / Pride is conquered in the company of the faithful; / There, in their midst, God's wisdom stands revealed. / God dwells near in the company of the faithful; / In the calmness of their presence all doubt is laid to rest. / There one obtains that precious jewel, the Name, / And striving by their aid one finds that blissful peace with God. / Who can hope to utter the wonder of their glory, / The glory of the pure and true in union with the Lord.

McLeod sees here that, after this definition of 'the company of the faithful' (*sangat, sat-sang* or *sadh-sang*), the next two octaves of the *Sukhmani* shift the focus to personal salvation, 'directing attention instead to a definition of the *braham-giani*' or self-realized person, although he does not spell out their effective complementarity. Incidentally, the Orientalist will also not allow the name of 'communion of saints', by which the early Christian church described itself, to describe any Sikh congregation, but we will let it pass.

The *braham-giani* is he who possesses an understanding of God's wisdom, the person who has found enlightenment in the company of the devout. Such a person acquires thereby an impressive range of virtues, some involving [*a*] his relationship with men and others [*b*] his relationship with God. The former includes such qualities as purity, humility, patience, kindness, and detachment. The latter pre-eminently requires [active] remembrance of the divine Name. He who devotes himself to the discharge of these obligations attains deliverance for himself and the power to confer it on others by means of word and example. Men teach and observe various beliefs concerning the means of deliverance. All should realise that there is but one way [the one tradition and the tradition of the one]. Evil is universally proscribed and the Name is accessible to all (1984: 110ff, 156).

I think that perhaps the penultimate octave of the *Sukhmani* (23), which I quote in Macauliffe's translation, sums up best the message of the-guru-in-the-scripture in relation to society, person and cosmos. It opens by explaining the mechanism of reflection whereby the individual and the collective are reconciled: 'In the company of the saints I have seen God within me.' It continues to explain that God does not dwell only in the heart of the subject: 'God dwelleth within as also without man.' He is the immanent, 'the whole creation is His body'; but he is also the transcendent, 'included in everything, He yet remaineth distinct'. He keeps throughout the full powers of transcendence: 'He blendeth with Himself whomsoever He pleaseth' and 'He to whom God showeth Himself, beholdeth Him'.

I infer that society and the person are reconciled ultimately as but two moments of the one cycle, the unity in the duality, ending as it began with the one, 'He Himself [performeth and] heareth His own praises'. Initially, even the appearance of duality has to be laid at his door, 'He hath made transmigration as a play, / And rendered Maya subservient to Him.'

He who knoweth Him must always be happy, / And God will blend him with Himself. / . . . That through him the Name might be remembered. / He was saved himself and he saved the world: / To him, Nanak, I ever make obeisance (transl. Macauliffe 1909: III, 266f).

The element of reciprocity, secondly, occurs and remains present and constant throughout in the relation of God and man: the-guru-in-the-scripture, transcendent and immanent, must be responsive to the-disciple-in-the-world. 'The perfect Guru fulfils all my needs,' says Guru Arjun, 'My enemies smitten, God's truth revealed.' Any individual person, group or congregation may consult the-guru-in-the-scripture in response to any problem at any time by 'taking a *hukam* (order)', which simply means duly opening the book of scripture at random and reading the first verse as it appears at the top of the left-hand page. This is in any case to be done publicly by every local congregation at the conclusion of the daily divine service, morning and evening. The reciprocal duty of God to man, including being the defence of the defenceless, is thus clearly stated in Guru Gobind Singh's address that is an obligatory part of the daily evening service (*Sodar rahiras*).

Extend to me your guiding hand, grant this my heart's desire, / That [my mind] at your feet, most gracious Lord, accepted I may dwell.

Let all my foes be overcome, your hand my sure defence. / Let all around me live in peace, all those within my care.

Your hand I crave, my rampart strong; destroy my foes this day. / May sweet [?] success crown all my hopes, my praise for you endure.

For you alone, Creator Lord, I follow and obey. / May all my people cross life's sea; may all my foes be slain.

Hold forth your hand when death draws near, let every fear depart. / Sustain me by your mighty strength, your sword [-banner] the sign I bear.

Protect me, Master, be my shield, with all who hold you dear; / The poor man's friend, the tyrant's foe, creation's only Lord . . . my Guru and my King.

The only door I seek, O Lord, is that which leads to you alone. / Hear me and keep me free from harm, save me your humble slave (*Benati chaupai* and *Dohara*, transl. McLeod 1984: 99f).

Guru-Khalsa

The study of religion has always suffered in modern times, among other things, from the dualism of eternity and time. Traditionalists, whether of the immanent or of the transcendent God, tend to place the most important features of 'religion-as-value' too far outside history and society, time and place. Marxist and liberal modernists, on the other hand, tend to see the most important features of 'religion-as-fact' wholly within the contexts of history and society. These two tendencies of course stand united against any relation of religion and modernity, except incidentally and contingently, because they both suppose that religion cannot produce its own modernity, as the Puritans apparently did, and also live to survive it, as the Commonwealth in England did not. If ever religion and modernity, whose true roots always lie in other factors, do happen to be found together, it is assumed to be a merely metonymic couple, of which religion is the one that must sooner or later adapt itself to the other partner.

But I think that any religion worth the name must begin and end with the non-dualism of fact and value, theory and practice, ends and means. Sikhism is therefore not unique but typical in its quest to find the point of conjunction between time and eternity; and its own internal modernity is to find in society the locus of self-realization, and not only of conformity to tradition or what is handed down or what is decreed by the prince, even to the point of having to exit this life in the name of God. All three relations that it typically seeks to establish among the trinity of Guru, Granth and Panth are relations of mutual participation and reciprocal embodiment through their point of trijunction which is mystically both the *sangat* and the *braham-giani*, society and the person. A congregation of the Khalsa anywhere and at any time is not simply an assembly of the like-minded and the devout. It is the locus of a compelling as well as transforming power or virtue, the Guru-Panth as the other half of the Guru-

Granth, and God can no more ignore it than he could forsake the man on the cross, because it/he is his own.

All Sikh ceremonies therefore consist of almost nothing but recitations of the scripture in the congregation, the word of God to man, followed by *Ardas*, the Sikh prayer, the supplication of man to God. The Sikh day is divided in the *gurudwara* or at home into two parts by such a simple ceremony and taking a *hukam* (order) for the next half-day. The minimum prescribed texts for individual recitation in the daily rule, e.g. with which children begin, are only three. (*a*) First thing early in the morning, Guru Nanak's *Japji*, preferably learnt by heart, to which an adult man or woman should add the *Jap* and the short *Ten savayyas* of the tenth guru, and the devout usually add the *Sukhmani* of the fifth guru. In a *gurudwara* the congregation, as far as I can see, comes individually prepared by the recitation of the *Japji* and joins in the liturgical opening of the Guru Granth Sahib with the same hymn (*Anand*) that will be used at the closing in the evening, or at any other time, and followed in the morning by *Asa di var*, a joint work of the first and the second gurus, set to music and sung as a community hymn. (*b*) In the evening at sunset, personally or in the *gurudwara*, the *Sodar rahiras*, made up of compositions of almost all the gurus who have left written memorials, but not arranged in chronological order. (*c*) *Kirtan sohila*, 'Sing to His glory', made up of brief compositions of the first, fourth and fifth gurus, prescribed for recitation, individual and collective, upon retiring for the night.

> When the appointed recitation or *kirtan* [recitation when set to music] draws to its close [for any ceremony] a portion of Guru Amar Das' *Anand Sahib* is read. *Ardas* is recited, a concluding hymn from the Guru Granth Sahib is read [as taking the *hukam*], and the service concludes with the distribution of [consecrated] *karah prasad* to all who are present. *Ardas* is also recited [by the daily rule] at the conclusion of early-morning and early-evening devotions (McLeod 1984: 103).

In fact, this would make the *Anand* of Guru Amar Das, the third guru, or strictly its first five and last verses, the most oft-repeated words of scripture, occurring in all ceremonies, including life-crisis rituals, and fieldwork will confirm this. It is certainly much loved by men and women, old and young, and it has even been said that to be a Sikh is to be in the state of *anand*, but I shall here quote only its first verse.

By the grace of the Eternal One, the true Guru:

When the Guru comes, O mother, joyous bliss (*anand*) is mine; / Boundless blessing, mystic rapture, rise within my soul. / Surging music, strains [hymns] of glory, fill my heart with joy; / Breaking forth in songs of gladness, praise to God within. / Comes the Guru, I have found him; joyous bliss is mine (transl. McLeod 1984: 100).

Apart from the scripture, the received and recorded word of the gurus, the only other works appointed by consensus to be read in *gurudwaras* as represented by the *Sikh rahit maryada* (the Sikh code of conduct, S.G.P.C. 1950) are the commentaries and compositions of Bhai Gurdas (*c.* 1550–1636), already mentioned, and Bhai Nand Lal (d. Multan 1712), who had been in the Mughal service before joining the tenth guru and so wrote in Persian. As Brother Gurdas says about the Sikh congregation as the society for salvation, 'The gathering of the faithful is the Realm of Truth (*sach khand*, the final stage or phase of self-realization and God's dwelling)' (*Var* 22: 18). He again repeats the principle that, on this path, society is to be preferred as sovereign when compared to the state, 'Our true King is the Guru; all the kings of the earth are false' (*Var* 15: 1). Brother Nand Lal's catechism of faith, wherein he personally addresses Guru Gobind Singh whose service he had joined late in life, definitively explains the interrelation of Guru, Granth and Panth.

The Catechism of Bhai Nand Lal

NAND LAL: You say that we should see you, O Master. Tell me where we are to find you.
THE GURU: Listen attentively, Nand Lal, and I shall explain. I am manifested in three ways: the formless or invisible (*nirgun*), the material or visible (*sargun*), and the divine Word (*gur-sabad*). The first of these transcends all that is material. It is the *neti neti* of the Vedas, the spirit which dwells in every heart as light permeates the water held in a vessel. The second is the sacred scripture, the Granth. This you must accept as an actual part of me, treating its letters as the hairs of my body. This truly is so.

Sikhs who wish to see the Guru will do so when they come to the Granth. He who is wise will bathe at dawn and humbly approach the sacred scripture. . . .

The Sikh himself is the third form which I take, that Sikh who is forever heedful of the words of sacred scripture (*gurbani*). He who loves and trusts the Word of the Guru is himself an ever-present manifestation of the Guru. Such a Sikh is the one who hears the Guru's words of wisdom and reads them so that others may hear. . . .

NAND LAL: You have told me of three forms, Master, the three being the invisible, the visible, and the Word of the Guru. The first of these we cannot see and the second we witness in the person of the obedient Sikh. Merciful Lord, how can we comprehend the infinity of your invisible form? You are the supreme Guru our Master, and your presence mystically pervades every soul, (but how can we perceive that presence)?

THE GURU: You are a devout Sikh, Nand Lal. Hear this divine message which I impart to you. First you must recognise the Guru as visibly present in his Sikhs and serve me by diligently serving them [*seva*]. Next you must serve me by singing the divine Word, accepting it as truly a manifestation (of the Guru). He who regularly sings portions of the sacred scripture shall thereby come to an understanding of the infinite being (of the Guru).

He who reads or hears this homily and pays careful heed to it will find himself the object of much admiration. He will also find that he is worthy to be united with me, his spirit mystically blended in mine.

This message of comfort and joy was delivered [three and a half years before the events of Anandpur] on the ninth day of the waxing moon in the month of Maghar, Samvat 1752 (4 December 1695 AD) (transl. from Ganda Singh's ed. by McLeod 1984: 76f).

Finally, the trinity of Guru, Granth and Panth can be seen and the purest accents of the Guru Panth or the Guru Khalsa can be heard manifest in the *Ardas*, the Sikh prayer of the congregation, standardized, collective and anonymous (*Sikh rahit maryada*, or code of conduct, S.G.P.C. 1950). According to the probably Parsi author of the *Dabistan-i mazahib*, written in Persian by someone who had observed Sikhism in the 1640s as a contemporary of the sixth and the seventh gurus, whenever anybody wanted a gift from heaven, he would approach such an assembly of Sikhs and ask them to pray for him. Indeed, 'Even the Guru asked his Sikhs to intercede for him' (Teja Singh 1951: 127).

The first part of the *Ardas* text, after calling the congregation to stand to order, consists of six lines of verse by Guru Gobind Singh himself invoking God and the first nine gurus, plus superadded reference by the community to Guru Gobind Singh as the tenth guru, and to the status of the Guru Granth Sahib. Then follows the body of the text, consisting of several paragraphs of rhythmic prose composed by successive generations of Sikhs 'as the events of their history unfolded', punctuated as indicated by the utterance, *Vahiguru*, by all who are present in the assembly (*sangat*). The third part, the supplicatory prayer for the occasion, can be and should be suitably varied by the person leading the congregation. The final couplet calls down God's blessing on everyone, Sikh and non-Sikh, and, like the opening lines, cannot be varied or omitted.

Resurrection of the Congregation: *Ardas*

Victory to the Lord, the Eternal One;
May Almighty God assist us.
The Tenth Master's Ode to Almighty God:

Having first remembered God, turn your thoughts to Guru Nanak; / Angad Guru, Amar Das, each with Ram Das grant us aid. / Arjan and Hargobind, think of them and Hari Rai. / Dwell on Siri Hari Krishan, he whose sight dispels all pain [sorrow]. / Think of Guru Tegh Bahadur; thus shall every treasure come [hastening into our homes]. / May they [all assist us everywhere,] grant their gracious guidance, help and strength in every place.

May the tenth Master [King], the revered Guru Gobind Singh [the lord of hosts and protector of the faith], also grant us "help and strength in every place". The light which shone from each of the ten Masters shines now from the sacred pages of the Guru Granth Sahib. Turn your thoughts to its message and call on God, saying, *Vahiguru*!

The Cherished Five [who were first invested with the collective guruship], the Master's four sons [two who d. martyrs, two who d. heroes, 1705], and the Forty Liberated [martyrs, who had strayed and returned to the fold]; all who were resolute, devout and strict in their self-denial; they [*shahid*, martyrs, and *murid*, disciples] who were faithful in their remembrance of the divine Name and [shared their wages or earnings] generous to others; they who were noble both in battle and in the practice of charity; they who magnanimously pardoned the faults of others: reflect on the [deeds] merits of these faithful servants [pure and true], O Khalsa, and call on God, saying, *Vahiguru*!

Those [*shahid*, martyrs, of both sexes,] loyal members of the Khalsa who gave their heads for their faith; who were hacked limb from limb, scalped, broken on the wheel, or sawn asunder [left from right for refusing forcible conversion]; who sacrificed their lives for the protection of hallowed *gurdwaras* never forsaking their faith; and who were steadfast in their loyalty to the uncut hair [unshorn being] of the true Sikh: reflect on their merits, O Khalsa, and call on God, saying, *Vahiguru*!

[Those who, to purge the temples of long-standing evils (at Nankana Sahib, 1921, and Guru-ka-Bagh, 1922), suffered themselves to be ruthlessly beaten or imprisoned, to be shot, cut up, or burnt alive with kerosene oil, but did not make any resistance or utter even a sigh of complaint: think of their patient faith and call on God!]

Remember the five *takhts* [lit. thrones, seats of temporal authority within the Panth] and all other *gurdwaras*. Reflect on their glory and call on God, saying, *Vahiguru*!

This is the first and foremost petition of the [whole] Khalsa, that God (*Vahiguru*) may dwell eternally in the thoughts of the entire [*sarbat*] Khalsa, and that by this remembrance all may be blessed with joyous peace. May God's [grace] favour and protection be extended to the Khalsa wherever its members may be found. Sustain it in battle, uphold it in the exercise of charity, and grant it victory in all its undertakings. May its name be exalted and may its enemies be subdued by the might of the sword [of God]. Call on God again, O Khalsa, repeating, *Vahiguru*!

Grant to your Sikhs a true knowledge of their faith, the blessing of uncut hair [being unshorn], guidance in conduct, spiritual perception, patient trust, abiding faith, and the supreme gift of the divine Name. May all bathe [*darshan, isnan*] in the sacred waters of Amritsar. May your blessing eternally repose on all who sing your praises, on the banners which proclaim your presence, on all places which provide shelter and sustenance to your people [or, may the Sikh choirs, banners, mansions ever abide]. Let us praise the way of truth [the kingdom of justice come] and call on God, saying, *Vahiguru*!

May the Sikhs be humble of heart yet sublime in understanding, their belief and honour committed to your care. [After the Partition of 1947, the following two sentences have replaced the paragraph in parenthesis quoted above in reference to the *gurudwara* movement, 1921.] O God, eternal Lord and Protector of the Panth, grant to the Khalsa continuing access to Nankana Sahib and to other *gurdwaras* from which it has been separated. Grant to its members the right to behold these sacred places and to care for them in the service of love. Merciful Lord, pride of the humble, strength of the weak, defence of the helpless, our true Father and our God, we [the local congregation] come before you praying that . . . [Refer here in appropriate words to the occasion for which the gathering or congregation has assembled and/or the names of the recitations completed.] Forgive us for any errors committed during the reading of the sacred scripture, and grant to all the fulfilment of their due tasks and responsibilities.

Bring us into the company of [only] those devout souls whose presence inspires remembrance of your divine Name.

Nanak prays that the Name may be magnified; / By your grace may all be blest (transl. McLeod 1984: 104f; interposed paragraph Teja Singh 1951: 123).

God's is the Khalsa; may God's be the victory!
God is truth!

A little elementary fieldwork would have assured the Orientalist that the published text is incomplete, since it does not write down its own frame, and that the ceremony of *Ardas* is not yet over. In fact, the congregation is now required to stand again, after bowing to the Guru Granth Sahib, and to all join in a final six-line chorus. McLeod has himself found the first two lines elsewhere in the code of conduct of Prahlad Singh, who claims to have personally received his instructions from the tenth guru, and also the last two lines in the code of conduct of Bhai Nand Lal (1984: 10, 78). The middle two lines simply reiterate the status of the Guru Granth Sahib; altogether there is no recent innovation here, therefore, and I have simply adapted and completed McLeod's own translation below by taking some help from Kapur Singh (1959: 449). I cannot say if the latter is right in supposing from his comparative studies that congregational prayer is 'an essentially Sikh institution in India'.

The Panth was founded at the command of *Akal Purakh* (the Eternal

One); / Every Sikh is bidden to accept the Granth as Guru. / Accept the Guru Granth as the manifest body of all the Gurus; / And the pure in heart shall find the true Word in it. / The Khalsa shall rule [Khalsa Raj] and no one shall defy it; / All who endure suffering and privation shall be brought to the safety of the Guru's protection.

Members of the congregation now resume their seats, the drum of sovereignty and self-rule is sounded, and everyone awaits in silence and with utmost expectancy the taking of the *hukam*, one's orders for the day or for the night to come. That is, the next half of the twenty-four hour cycle must be commenced with this taking of the *hukam* in the liminal moment before the past half is considered closed. Only after that, and the distribution of *karah prasad* to all, may the congregation disassemble and persons individually take their leave from the *gurudwara* and from one another with the same final words of amen as before.

And so we too should take our leave of the story of Sikhism and the Indian modernity, each of them in relation to itself, the world and the other. We did not discover here any plan, blueprint or system, but the method and the praxis, theoretical and operational, of non-dualism, freedom and responsibility when addressed equally to religion, the state and civil society. Sikhism is not a tradition handed down at all but an experiment in the sense of Gandhi's autobiography, *The story of my experiments with truth*, the conjoint experiment of Guru, Granth and Panth. Perhaps this method of non-dualism is the secret lever of motion within Indian culture, history and society as well as the self of religion.

Yet Sikhism is not complete nor self-contained; it needs to be met with pluralism, mediation of the one and the many, in civil society even more than secularism in the state. What is the contribution of Sikhism in relation to the Indian theatre of civil society?

Our new semiological analysis of its known history, 1500 to 1947, shows Sikhism to consist, not of a syncretism, but of three united problems, the theory of the name, the story of martyrdom versus kingdom and the institution of a society for salvation. These three aspects may have unfolded in sequence corresponding to the emphases respectively of the first guru, the fifth guru and the tenth guru, but it is their mutual coherence that was highlighted here in the study of symbol and text, context and event. Together the guru and the Sikh established a pattern of the Indian modernity in religion as well as civil society versus the state, and to show this it is necessary to first explain the underlying structure of medievalism as it was before 1500, in the Hindu culture as well as the Muslim culture. Finally, an examination of the period of collaboration between Sikhism and Gandhism, viewed now after the new revolution of Islam in Iran, argues for a recognition among students of human culture of the figure of the martyr, rather than of the hero or the victim, as the universal foundation of civil society.

Somewhere between the *atma* and *Paramatma*, the soul and the All or the Lord, the orginal self-sacrifice of the martyr introduced both history and society in effect as new forms, aspects or prospects of non-dualism, whether in Sikhism or in the project of some other Indian modernity. This then is the space of pluralism and civil society, or of self, the world and the other, opened by the exemplars between the priest as custodian of tradition, on the one hand, and the prince, maker or executor of the state, on the other, and often in conflict with both of them.

References Cited

ABDUL MAJID KHAN

1967a. The impact of Islam on Sikhism. *In* Simla 1967: 219–29.

1967b. Muslim devotees of Guru Gobind Singh. *In* Chandigarh 1967: 89–92.

BELLAH, ROBERT N.

1964. Religious evolution. *American sociological review* (Albany, New York), 29: 358–74.

CAMBRIDGE

1970. *The Cambridge history of Islam*, 2 vols, ed. Holt, P.M., Ann Lambton and Bernard Lewis. Cambridge: University Press.

CHANDIGARH

1967. *The tenth master. Tributes on tercentenary*. Chandigarh: Guru Gobind Singh Foundation.

CORBIN, HENRY

1970. *Creative imagination in the Sufism of Ibn Arabi*. London: Routledge and Kegan Paul.

DE BARY, W.T. *et al.*, eds

1958. *Sources of Indian tradition*. New York: Columbia University Press.

DUMONT, LOUIS

1959. A structural definition of a folk deity. *Contributions to Indian sociology* (Paris), 3 (O.S.): 75–87.

1960. World renunciation in Indian religions. *Contributions to Indian sociology* (Paris), 4 (O.S.): 33–62.

1962. The conception of kingship in ancient India. *Contributions to Indian sociology* (Paris), 6 (O.S.): 48–77.

FAZLUR RAHMAN

1970. Revival and reform in Islam. *In* Cambridge 1970: II, 632–56.

GANDA SINGH

1975. Guru Nanak's impact on history. *In* Patiala 1975: 418–27.

GANDHI, M.K.

C.W. The collected works of Mahatma Gandhi, 90 vols, index, suppl. New Delhi: Publications Division, Ministry of Information and Broadcasting, 1958–84, 1988, 1991.

GLASSÉ, CYRIL

1989. *The concise encyclopaedia of Islam.* London: Stacey International.

GONDA, JAN

1965. *Change and continuity in Indian religion.* The Hague: Mouton.

GURCHARAN SINGH

1976. City of the golden temple. *Illustrated weekly of India* (Bombay), 97 (52), 26 December 1976: 40–7.

GURMUKH NIHAL SINGH, ed.

1969. *Guru Nanak. His life, time and teachings.* Delhi: National Publishing House *for* Guru Nanak Foundation.

HABIB, MOHAMMAD

1961. Introduction. *In* Nizami 1961: iii–xxii.

HARDY, PETER

1958. Islam in medieval India. *In* de Bary 1958: 367–528.

HUGHES, T.P.

1895. *Dictionary of Islam.* London: W.H. Allen.

INDUBHUSAN BANERJEE

1962. *Evolution of the Khalsa*, vol. 2. *The reformation.* Calcutta: A. Mukherjee and Co. (First published in 1947).

1963. *Evolution of the Khalsa*, vol. 1. *The foundation of the Sikh Panth.* Calcutta: A. Mukherjee and Co. (First published in 1936).

JAHANGIR

1909. *Tuzuk-i Jahangiri. Memoirs of Jahangir*, vol. 1, transl. and ed. Rogers, A. and H. Beveridge. London: Royal Asiatic Society.

KAPUR SINGH

1959. *Parasharprasna, or the Baisakhi of Guru Gobind Singh.* Jullundur: Hind Publishers.

KEER, DHANANJAY
1950. *Savarkar and his times.* Bombay: A.V. Keer.

KNOTT, JOHN R.
1993. *Discourses of martyrdom in English literature, 1563–1694.* Cambridge: University Press.

LEVI-STRAUSS, CLAUDE
1977. *Structural anthropology*, vol. 2. London: Allen Lane.

LEVY, REUBEN
1957. *The social structure of Islam.* Cambridge: University Press.

LOEHLIN, C.H.
1967. Guru Gobind Singh and Islam. *In* Simla 1967: 109–18.

LOTMAN, J.M. *et al.*
1975. Theses on the semiotic study of culture. *In* Sebeok, T.A., ed., *The tell-tale sign. A survey of semiotics*, pp. 57–83. Lisse, Netherlands: Peter de Ridder Press.

MACAULIFFE, M.A.
1909. *The Sikh religion. Its gurus, sacred writings and authors*, 6 vols. Oxford: Clarendon Press.

MCLEOD, W.H., transl. and ed.
1984. *Textual sources for the study of Sikhism.* Manchester: University Press.

MARX, KARL
1843. On the Jewish question. *In* Marx, K. and F. Engels, *Collected works*, vol. 3, pp. 146–74. Moscow: Progress Publishers, 1975.

MOHAN SINGH DIWANA
1967. Arabic-Persian key words in Sikhism. *In* Simla 1967: 249–59.

MOMEN, MOOJAN
1985. *An introduction to Shi'i Islam.* New Haven and London: Yale University Press.

MUJEEB, M.
1969. Guru Nanak's religion, Islam and Sufism. *In* Gurmukh Nihal Singh 1969: 114–21.

NIHARRANJAN RAY
1967. Introductory address. *In* Simla 1967: 1–12.
1970. *The Sikh gurus and the Sikh society. A study in social analysis.* Patiala: Punjabi University.
1975. The concept of Sahaj in Guru Nanak's theology. *In* Patiala 1975: 57–70.

NIZAMI, KHALIQ AHMAD
1961. *Some aspects of religion and politics in India during the thirteenth century.* Bombay: Asia.

PANDEYA, R.C.
1975. The philosophy of name. *In* Patiala 1975: 76–82.

PANDIT, P.B.
1977. *Language in a plural society.* Delhi: Dev Raj Chanana Memorial Lectures/Delhi University Press.

PARTHA N. MUKHERJI
1984. Gandhi, Akalis and non-violence. *Man and development* (Chandigarh), 6 (3): 58–77.

PATIALA
1975. *Perspectives on Guru Nanak*, ed. Harbans Singh. Patiala: Punjabi University.

PINCOTT, FREDERIC
1895. Sikhism. *In* Hughes 1895: 583–94.

POCOCK, DAVID F.
1960. Sociologies—urban and rural. *Contributions to Indian sociology* (Paris), 4 (O.S.): 63–81.

QURESHI, I.H.
1970a. Muslim India before the Mughals. *In* Cambridge 1970: II, 3–34.
1970b. India under the Mughals. *In* Cambridge 1970: II, 35–63.

RIZVI, S.A.A.
1975. Indian Sufism and Guru Nanak. *In* Patiala 1975: 191–222.

ROSE, H.A., compiler
1911, 1914. *Glossary of tribes and castes of the Punjab and north-west frontier province*, vols 2, 3. Lahore: Samuel T. Weston/Civil and Military Gazette Press.

SAVARKAR, V.D. (pseud., A Maratha)
1923. *Hindutva.* Nagpur: V.V. Kelkar.

SCHOLEM, GERSHOM G.
1965. *On the Kabbalah and its symbolism.* New York: Schocken Books.

SCHUON, FRITHJOF
1963. *Understanding Islam.* London: Allen and Unwin.

SGPC (Shiromani Gurdwara Prabandhak Committee)
1990. *Sikh rahit maryada* (Sikh code of conduct, in Punjabi). Amritsar: S.G.P.C. (First published in 1950).

SHARIATI, ALI
N.D. *Martyrdom* (*shahadat*), transl. Bakhtiar, Laleh and Husayn Salih. Teheran: Abu Dharr Foundation.

SIMLA
1967. *Sikhism and Indian society. Transactions of the Indian institute of advanced study*, 4. Simla: Indian Institute of Advanced Study.

TEJA SINGH
1951. *Sikhism. Its ideals and institutions*, 2nd ed. Bombay, Calcutta, Madras: Orient Longmans. (First ed. published in 1938).

UBEROI, J.P.S.
1967. On being unshorn. *In* Simla 1967: 87–100.
1972. The elementary structure of medievalism. Paper presented at Indo-French seminar on anthropological linguistics. Patiala: Punjabi University.
1984. Sikhism and Islam. *In* Amita Ray, H. Sanyal and S.C. Ray, eds, *Indian studies. Essays presented in memory of Niharranjan Ray*, pp. 223–30. Delhi: Caxton Publications.

UBEROI, PATRICIA and J.P.S. UBEROI
1976. Towards a new sociolinguistics. *Economic and political weekly* (Bombay), 11 (7): 637–43.

VAN GENNEP, ARNOLD
1960. *The rites of passage*. London: Routledge and Kegan Paul.

WEBER, MAX
1948. *From Max Weber. Essays in sociology*, transl. and ed. Gerth, H.H. and C. Wright Mills. London: Routledge and Kegan Paul.

WENSINCK, A.J.
1922. The Oriental doctrine of the martyrs. *Mededeelingen der koninklijke akademie van wetenschappen* (Amsterdam), series A, 53: 147–74.

ZAKIR HUSAIN
1969. Foreword. *In* Gurmukh Nihal Singh 1969: v–vi.

Index